The Audience in Everyday Life

The Audience in Everyday Life
Living in a Media World

S. Elizabeth Bird

Routledge
New York and London

Published in 2003 by
Routledge
29 West 35th Street
New York, NY 10001
www.routledge-ny.com

Published in Great Britain by
Routledge
11 New Fetter Lane
London EC4P 4EE
www.routledge.co.uk

Routledge is an imprint of the Taylor & Francis Group.
Printed in the United States of America on acid-free paper.

10   9   8   7   6   5   4   3   2

Chapter 2 is an extensive revision of "What a Story! Understanding the Audience for
Scandal," first published in *Media Scandals: Private Desire in the Popular Culture
Marketplace*, edited by J. Lull and S. Hinerman (Cambridge, U.K.: Polity Press, 1997).
Chapter 3 is an extensive revision of "Chatting on Cynthia's Porch: Building Community in
an Internet Fan Culture," first published in *Southern Communication Journal* 65:1, 49–65,
1999. Chapter 4 is an extensive revision of "'Indians are like that': Negotiating identity in a
Media World," first published in *Black Marks: Minority Ethnic Audiences and Media*, edited
by K. Ross and P. Playdon (Aldershot, U.K.: Ashgate, 2001). Chapter 6 is an extensive
revision of "CJ's revenge: Media, Folklore, and the World of AIDS," first published in *Critical
Studies in Mass Communication* 13, 1–15, 1996.

Library of Congress Cataloging-in-Publication Data

Bird, S. Elizabeth.
   Living in the media world / by S. Elizabeth Bird.
      p.   cm.
Includes bibliographical references and index.
   ISBN 0-415-94259-4 (hardback : alk. paper)—ISBN 0-415-94258-6 (pbk. : alk. paper)
   1. Mass media—Social aspects.   2. Mass media—Audiences.
3. Mass media and culture.   I. Title.
HM1206.B57   2003
302.23–dc21   2003011583

ISBN 0-4159-4259-4(hb)
ISBN 0-4159-4258-6(pb)

For Graham, with love

CONTENTS

ACKNOWLEDGMENTS

This book brings together projects I have worked on for several years, both in Minnesota and Florida, and many people have contributed, both directly and indirectly.

I would like to acknowledge the support of the University of Minnesota Graduate School Grant-in-Aid program, which provided funding for the studies that led to chapters 2 and 4, as well as the College of Liberal Arts at the University of Minnesota, Duluth, which provided small funding opportunities that supported my ethnographic research.

My work has been greatly facilitated by hard-working graduate assistants, most notably David Woodward in Minnesota, and recently Noah Porter, who efficiently carried out essential bibliographic research as I worked on this manuscript at the University of South Florida.

I am especially appreciative of the sharp and open minds of the Anthropology and Communication graduate students in my USF seminars on Anthropology, Media, and Contemporary Culture, and Visual Anthropology. Their critical participation has helped tremendously as I have tried to apply an interdisciplinary cultural perspective to the thorny issues of media's role in our culture.

Many anonymous reviewers have commented on various versions of shorter papers that became some of the chapters in this book, and I believe the final manuscript owes a great debt to them all.

Many people have willingly participated in my research over the years, being interviewed, taking part in focus groups, and allowing me into their homes and lives. In particular, I thank the members of the DQMW-L list, who not only allowed me into their community, but actively helped shape my research through their insightful comments and suggestions.

I am lucky to have fine colleagues and friends in the Departments of Anthropology and Communication at USF, who have offered both professional insights and personal support, especially Jane Jorgenson, Linda Whiteford, Gil Rodman, Brent Weisman, Mark Neumann, Susan Greenbaum, Fred Steier, and Michael Angrosino. And I especially thank my colleague and friend Bob Dardenne, of USF-St. Petersburg. We began talking about media and culture almost twenty years ago, and we haven't stopped since.

Deep thanks go to my dearest friend, Eve Browning. Minnesota is far from Florida, and our daily walks are over. But through the wonder of e-mail, we can still talk, laugh and bitch at the world, and that helps keep me sane.

Finally I thank my families. My birth family—my father, who always believes in me, my mother who would have loved to see books with my name on them, and my siblings Michael, Alan, and Alison, who don't always know what I do, but are unfailingly encouraging. And above all I thank the family who has to live with my strange enthusiasms. My sons, Tom and Dan, fill me every day with love and pride, and are everything any mother could dream. Last but never least, the man whose love and support have been constant for so many years, from England to Iowa, to Minnesota, and now to the surreal world of Florida—my husband, Graham Tobin. Thank you, with love.

1

BEYOND THE AUDIENCE
Living in a Media World

INTRODUCTION: MEDIA AS CULTURE

Even as everyone appears to acknowledge Western culture's "media-saturated" reality, many of us deny it in our own lives. As a media scholar working within the discipline of anthropology, I get used to the comments: "I make up my own mind about politics; I don't need the media"; "I guess what you do is interesting, but of course I never watch TV myself." Many anthropology students resist the idea that studying media is relevant to an understanding of contemporary culture. Interesting, perhaps— an elective that offers diversion from the serious pursuit of their discipline—but hardly significant. Yes, media messages do insidious things to people, but not of course to me.

I think the common sense view I encounter so frequently in my discipline is symptomatic of the way many Americans think of the media. They are aware that images are everywhere, and they sometimes feel beleaguered by the onslaught of messages. They worry about the "effect" of media on children, and they seethe about "biased" reporting or negative stereotypes. Yet they still resist full acknowledgement that today the media are not extrinsic to Western culture, but fully internalized—that "the obvious but hard-to-grasp truth is that living with the media is today one of the main things Americans and many other human beings do" (Gitlin 2001, 5).

Each time I teach a graduate seminar on the media in contemporary culture, I begin with the time-honored "media deprivation exercise" (Dardenne 1994; Mastrolia 1997), asking students to avoid all media for

1

a week, and to write a journal about their experiences. At the outset, most are blasé; they hold images of themselves as serious, committed individuals, living lives free of external influences. Some proudly point out that they do not own a television. A week later they return, some almost shamefaced, as their journals and class discussion take them in directions they had never expected. A mother talks about how she is accustomed to watching a particular TV program with her teenage daughter, and then talking about and "around" it in a comfortable moment of connection. This week she had missed that terribly, and her daughter was irritable and upset by the break in routine. The issue was not the message of the show itself, but the relational context they had built around it. Another describes vividly his genuine fear when he had to turn off the radio for his daily 45-minute commute. "Suddenly there were all these sights and sounds that seemed to overwhelm me, and I felt disoriented. Later I actually heard a birdcall!" Social lives are put on hold, because they included movies or music. A political junkie feels withdrawal from news, and realizes she cannot enjoy her continuous processing of current events without media input. Another acknowledges her dependence on regular visits to an Internet chat forum. No one lasts more than four days; it is just too difficult. "It felt like withdrawing from the world," writes one ruefully. At this point, we are ready to start looking seriously at the media *as* culture.

A striking point that emerges from the deprivation experience is that, although it tends to confirm the cliché of "media saturation," it also shows that as individuals, we experience media in non-predictable and non-uniform ways. One can be a proud TV-avoider, yet still be almost physically dependent on recorded music. One can watch TV most of the time as a casual, passive viewer, but be a knowledgeable, active "fan" of a particular program. The images and messages wash over us, but most leave little trace, unless they resonate, even for a moment, with something in our personal or cultural experience.

The amorphous nature of media experience was brought home to me when a doctoral student in my department almost reluctantly approached me to discuss themes emerging from her dissertation research, which was an analysis of the narratives of women who had been sexually abused while in the military (Redvern-Vance 1999). She was discovering that as women told their life stories, they would constantly use media references to interpret their experiences. These references ran the gamut from news, to talk shows, to soap operas. They invoked names like Oprah, Bill and Hillary, Monica, Anita Hill, Kelly Flynn (the Air Force pilot discharged for adultery and lying in a notorious 1997 case). One talked about her childhood identification with Popeye, Mighty Mouse and Superman.

One woman's "hero" was Archie Bunker, because "he let people know that white people can be ignorant"; later she identified with the film *In the Heat of the Night,* and named her dog Mr. Tibbs. Another, unable to articulate the depths of her emotions, lent the researcher videos, such as the film *Priest,* which focuses on sexual abuse, explaining that these would help tell her story. Yet another scheduled her interviews around her TV-watching obligations, and quite explicitly defined her own life as a "soap opera."

This student, a perceptive researcher, realized it had never occurred to her to ask specifically about media; certainly she was not doing what we might call "audience research." Yet eventually she could not escape the reality of media as a kind of cultural frame that pervaded her respondents' lives. Her understanding of that led her to a greater ability to relate to the women with the empathy needed for the sensitive research she was doing.

All this leads me to a discussion of why the notion of the "audience" has become so problematic in media studies. We really cannot isolate the role of the media in culture, because the media are firmly anchored into the web of culture, although articulated by individuals in different ways. We cannot say that the "audience" for Superman will behave in a particular way because of the "effect" of a particular message; we cannot know who will use Superman as some kind of personal reference point, or how that will take place. The "audience" is everywhere and nowhere. In this book I try to explore this rather vague notion of how we actually interact with the media. In so doing, I hope to contribute both to the ongoing debate on "the audience" within media studies, and also to suggest to fellow anthropologists how important it is to come to grips with the central role of media, both in contemporary Western culture and in rapidly-changing cultures all over the world. My case studies focus on people as active, selective makers of meaning; our culture may be "media-saturated," but as individuals we are not, or at least not in any predictable, uniform way. In my final chapter, I return to a question that rightly dogs us active audience advocates–but what about power and constraint? We may be able to make creative, individual meanings from this torrent of messages and images, but we can still only work with what we're given.

WHAT ABOUT THE AUDIENCE?

During the late 1980s, a flurry of scholarly activity effectively dismantled the idea that there really can be an "audience" out there waiting to be studied (e.g., Allor 1988; Erni 1989; Grossberg 1988; Fiske 1988;

Radway 1988). It has been established that the very conception of the word tends to reify the "transmission" view of communication, whereby a message is transmitted to a receiver, with varying degrees of interference or "noise" affecting the impact of the message. And, as Hartley (1992) and others have pointed out, it separates "them"–the audience– from "us"–the researchers, in a way that belies the reality that all of us are living in a mediated culture. Most people play many roles in their lives, and "being an audience" is probably not that important a one. The audience is an everchanging, fluid concept: "The conditions and boundaries of audiencehood are inherently unstable" (Moores 1993, 2).

So in understanding the role of the media in contemporary Western culture, we must somehow grasp the quality of the kaleidoscope, exploring how media articulate with such factors as class, gender, race, leisure and work habits, and countless other variables. The problem of course is that once we accept this pervasive, culturally-embedded nature of the media, it becomes very difficult to conceive of studying the phenomenon, without falling into the trap mentioned by Radway (1988) in which "Users are cordoned off for study and therefore defined as particular kinds of subjects by virtue of their use not only of a single medium but of a single genre as well" (1988, 363). As Seiter (1999) asks, "How do we draw the line in our data collection between audience research and the study of society, the family, the community?" (p. 9).

Radway saw ethnography as the only solution, advocating a major collaborative project in which a team of ethnographers fan out across an entire city and study how people's leisure practices are articulated in their lives. She does agree that this could be "potentially unwieldy and unending"(p. 369)—an understatement, to say the least. This kind of comprehensive endeavor has not been taken up as a serious project. Indeed, for some years now, audience scholarship has had something of a dilemma. The text-based response studies are seen as inadequate in capturing the kaleidoscopic quality of our media culture; if we cannot define an audience, is it effectively impossible to study it? Furthermore, the postmodernist "crisis" in anthropological representation (Clifford and Marcus 1986) left many uncertain about whether it is even valid to attempt to "speak for the other," making ethnography itself problematic (Bird 1992b). It seemed as though media reception studies were in something of a funk. Is our only option simply to hang out and wait for people to mention media in the course of everyday conversations and actions, or in interviews about other subjects? In that case, we will tend to produce variously rich anecdotes, like Redvern-Vance's inadvertent stories, or like Barker's (1998) discussion of his friends' conversations about movies, in which he perceptively calls into question the very terminology of audience research—"exposure," "arousal," "watch."

THE OPENING UP OF ETHNOGRAPHY

As Alasuutari (1999) writes, we are now seeing a "third generation," of reception studies, building on the models represented by Stuart Hall's encoding/decoding approach and the now-classic qualitative audience studies of Ang (1985), Lull (1990), Morley (1980, 1986), Radway (1984), Hobson (1982), and so on, but moving in rather broader directions.[1] And we are seeing anthropologists entering the field of media reception studies, meeting and engaging with the anthropologically-inspired work that in the past has largely originated from within the fields of cultural studies and communication. As recently as 1993, Spitulnik argued that "There is as yet no 'anthropology of mass media'" (p. 293); today we are finally beginning to see a body of literature that defines itself as just that (e.g., Askew and Wilk 2002, Ginsburg, Abu-Lughod, & Larkin 2002).

Essentially, this interdisciplinary "third generation" approach acknowledges the very real problems associated with trying to separate text/audience from the culture in which they are embedded, yet also accepts that it may be perfectly valid to enter the discussion through one particular genre or medium. Alasuutari (1999) argues that a "third generation" approach does not necessarily abandon specific audience studies, but that "the objective is to get a grasp of our contemporary 'media culture,' particularly as it can be seen in the role of the media in everyday life ... " (p. 6). Thus the goal must be to contextualize and to draw connections between media/audience and the larger culture. On the one hand, this opens the door to cast our net wider—to explore the kind of "opportunistic ethnography" that shows us "media culture" in action in a legitimate and fruitful way, most usually in the culture within which the researcher is already comfortable. For instance, Barker's (1998) account of his friends' response to a movie is not unlike an episode from the accounts of anthropologists who "hang out" over an extended period of time in another culture, observing carefully and keeping systematic notes over time (Emerson, Fretz, & Shaw 1995). His analysis of how one young man apparently used the character of Judge Dredd to frame his sense of identity as a "fascist" (Barker 1997) is another example of a similar approach, while Couldry (2000) is continuing to develop the intriguing notion of a "passing ethnography," which takes into account the mobile, ephemeral nature of much social interaction today. Anthropologist John

Caughey was a pioneer in media anthropology with his exploration of people's "imaginary" relationships with media figures. His studies go far beyond any notion of a fixed, static "audience" (Caughey 1984, 1994). Rather, they speak volumes about how media images are naturalized into everyday American life, just as spiritual and mythological images are naturalized in oral cultures. His nuanced, sensitive account of one woman's "imaginary relationship" with actor Steven Segal is an intriguing exploration of how significant media can be in both personal identity and enculturation (Caughey 1994). Occasionally, other anthropologists have followed this lead, focusing not on specific text/audience relationships, but on everyday experience, as in Fisherkeller's provocative study of adolescent identity construction, in which she uses an in-depth life-history approach to conclude that teenagers "look to television culture, consciously or not, to acquire *imaginative strategies* for acting on their dreams and hopes for the future, and for coping with social dilemmas" (1997, 485, italics in original). These kinds of analyses, focused on small-scale explorations of individuals' processes of meaning-making, can take us in directions that are quite different from conceptions of the static audience.

At the same time, we do not need to close the door on systematic, ethnographically-inspired studies that seek to explore specific moments of media interaction, as for example Seiter (1999) does with television. Like many before him, Alasuutari espouses an ethnographic approach as the ideal. However, he suggests that we might do well to follow the lead of anthropologists and move beyond a definition of ethnography that equates it with long-term participant-observation. As he writes, "It has been argued that a proper ethnographic study in audience ethnography entails at least several months stay in the 'field'" (1999, 5). From this perspective, the qualitative reception studies done over the years are failures as ethnography, based as they are mostly on interviews. Evans (1990), along with Wester and Jankowski (1991) argues that many audience studies touted as ethnographic do not meet the requirements of true ethnography, while Murdock (1997) decries the tendency in cultural studies to "stretch the definition of ethnography to cover almost any effort to collect extended accounts of people's beliefs, responses and experiences" (p. 184). He argues that such techniques as interviews, focus groups, diaries and so on "can offer a solid basis for creative interpretation—but...cannot provide thick descriptions" (p. 184) because they lack "a rounded account of the ways that people's utterances, expressions and self-presentations are shaped and altered by the multiple social contexts they have to navigate the course of their daily lives" (184–5). While advocating a true "return to ethnography" (p. 185), Murdock gives no guidance as to how, for example, a media reception study could realistically achieve the ideal of holism he seems to be advocating. Maybe ethnography really is impossible after all, as some have seriously argued (see Bird 1992b).

Yet perhaps ironically, this search for a pure ethnography comes "at a time when anthropologists...are increasingly questioning the whole notion of a 'field'" (Alasuutari 1999, 5) and wondering about the ideal of holism (Marcus 1986). Furthermore, as any anthropologist knows, ethnography has also long encompassed a range of methods that supplement or even replace classic fieldwork, and that may perfectly legitimately be used to study media reception–especially if combined with a broader analysis of cultural context. "Ethnography is not, in and of itself, a way to gather information" (Angrosino 2002, 3). Texts for student anthropologists outline many methods and approaches, including life histories, autobiographies, personal narratives, self-descriptions, diaries, interviews, and (dare I say it!) surveys and other quantitative methods (Bernard 1998; Angrosino 2002). The message from anthropology these days is that methods should be chosen not by using some standard of ethnographic "purity," but because they are appropriate for a particular project.

Classic ethnographic fieldwork may not be an appropriate method for studying dispersed media audiences, at least for the ethnographer working alone. Traditional participant-observation is most obviously necessary when studying a culture that is completely foreign to the researcher—when a new language must be learned or an unfamiliar worldview understood. Media anthropologists like Miller (1992), Wilk (1994), or Hahn (1994) quite rightly spent considerable time in the field. It may also be extremely valuable in contemporary society, when studying relatively self-contained sub-cultures and institutions, especially those with developed social conventions and rules. But many media ethnographers are studying cultural phenomena with which they are already familiar as participants, and there is a variety of techniques that build on this familiarity. In these situations "'fieldwork' has actually started years before" (Alasuutari 1999, 8), allowing the native researcher to focus attention on phenomena that are already quite familiar, whether the context is India (Parameswaran 1999; Mankekar 1999) or the United States. And we should not agonize unnecessarily about pure "holism" as a goal. Few anthropologists study complete, self-contained societies any more (if they ever did), but write ethnographies that explore specific questions and issues. Their holistic perspective emerges in the attempt to see these questions and issues in context, and linked to other aspects of the culture; in that regard holism is still an important anthropological credo, even as we understand that it is no longer a pure reality in an interconnected world.

WHY METHODS MATTER

Rather than worry about relatively insignificant issues like time spent in the field, I believe we should be thinking more carefully about matching suitable methods to the subtle questions we are trying to ask. Our aim should be to achieve an "ethnographic way of seeing," (Wolcott 1999) whose goal Emerson, Fretz, and Shaw (1995) define as "to get close to those studied as a way of understanding what their experiences and activities *mean to them*" (p. 12, italics in original). The chapters in this book represent my efforts over the years to investigate the role of the media in everyday U.S. culture, using a variety of methodological approaches, but guided by that way of seeing. I aim to shed light on how people interact with media to create meaning in their everyday lives. I agree with Morley (1999) that in spite of our understanding of the complexity of media's role in culture, we should not abandon case studies: "Rather, one reframes and recontextualizes them in a new way" (Morley, 1999, p. 196). Unlike many researchers before me, I am not focusing on one particular genre such as television, or separating factual from fictional media, but am trying to look across the spectrum of culture, stopping to highlight particular moments of cultural media interaction.

Classic "encoding/decoding" audience studies, with their usual use of focused, directed questions for artificially-constructed groups, are indeed limited in their ability to evoke the broader cultural context I have been discussing. For that reason, I have attempted to widen my range of approaches, experimenting with different methodologies as I try to tease out the way the media actually function in everyday life. As Nightingale writes, ethnographic studies strive with various degrees of success to move past the direct text/audience relationship: "The text, as work, has a finite quality . . . But there is another text, just as important but infinitely more elusive. It is the text which lives in the community of its users and which 'enters into life.' This is the text that I believe the cultural studies audience experiment tried to capture" (1996, 107). This "text" is still the one I am trying to capture as I explore everyday meaning-making through media.

And I believe our methodological choices matter in this respect. They matter, not because I subscribe to the "comfortable assumption that it is the reliability and accuracy of the methodologies being used that will ascertain the validity of the outcomes of the research, thereby reducing the researcher's responsibility to a technical matter" (Ang 1996, 47).

As Ang points out, that traditional view of social research implies that if somehow the researcher gets the methods right, then the results are necessarily and transparently true. As she goes on to explain, we can no longer legitimately accept that argument, the more we learn about the dynamic between researcher, method, and conclusions. In fact, methods matter because the choices made, along with the very characteristics of the researcher, play into and ultimately shape the conclusions of any research. And so it behooves us to examine our choices, and at the very least reflect on that dynamic interplay. Are we asking the right questions? Are our approaches consistent with the communicative styles of the people with whom we speak? Is the structured nature of our interaction driving responses in particular directions? Are we allowing our participants opportunities to define and direct the flow of communication? In working with graduate student researchers over the years, I have tried to push them to examine their methodological choices more carefully. Is the fact that they report very individualistic responses connected with their choice of one-on-one interviews? In what situations might observation yield different information from direct questioning? How does their personal identity affect the very nature of the ethnographic encounter?

Anthropologists have learned that many years of living with a culture does not necessarily ensure understanding or communication, at least not with everyone in the culture studied. For example, Keesing (1985) discusses the inevitable way methodology and theory intertwine in the ethnographic study of gender. He points out that in many ethnographic accounts, women appear silent, allowing ethnographers to conclude that women actually are "muted" within their culture. He himself studied a Melanesian culture, the Kwaio, for 20 years, and was being persuaded that women were largely non-reflective and silent. However, he then began using new methods within new contexts to try to encourage women's autobiographical reflections, building on his intimate knowledge of the culture and adding the special insight of a woman ethnographer. Silence was "transformed into an amazingly rich corpus of articulate accounts by a dozen women . . . " (p. 27).

With admirable humility, he concluded that the perceived silence of women was really an artifact of a 20-year problem with his ethnographic method, even though he was following standard participant-observer strategy. He notes: "I argue that (a) neither 'muteness' nor articulate accounts of self and society represent a direct reflection of 'women's status' or the role of women in society; rather (b) what women can and will say is a product of specific historical circumstances; and (c) emerges in a specific micropolitical context both of male-female relations and the ethnographic encounter itself; therefore that (d) whatever texts derive from such an encounter . . . must be interpreted in terms of these historical circumstances and micropolitics, which inextricably include the ethnographer her/himself . . . 'muteness' must always be historically and contextually situated, and bracketed with doubt" (p. 27).

INTERROGATING THE ETHNOGRAPHIC ENCOUNTER

One need hardly add that this applies to respondents of any gender, ethnicity or other identity. So at this point I wish to reflect a little on how our choice of methodological approaches might affect our ability to collaborate effectively with our participants in producing ethnographic understanding of media reception, using my own studies as examples. As Keesing suggests, I wish to focus attention on the "ethnographic encounter" itself, as an act of communication that is inseparable from the existing social and cultural circumstances of which it is part. In my discussion of the implications of particular methodological choices, my intention is not to offer a systematic run-through of all the methods used in this book. Rather, I want to focus attention on some of the approaches I have used in various ethnographic projects, reflecting in some detail on some of them, and suggesting that a more careful and self-reflexive considering of methods will serve us well as we move toward a more richly contextual understanding of "audience" behavior.

There are, of course, many kinds of ethnographic encounter. For instance, many feminist and critical audience researchers, in attempting to replace social science methodology, have embraced the face-to-face, in-depth interview as a valuable road toward experiential understanding (Langellier and Hall 1989). In fact, as van Zoonen (1994) points out, "ethnographic" audience studies have relied largely on this method, with a few forays into group interviews. According to Langellier and Hall, what is often lacking is detailed discussion of the interview process itself. As they write, "The interview is itself a communicative event with particular norms and rules. Therefore, researchers must critically examine the compatibility of the interview with the communicative norms of the interviewees" (p. 202)—precisely the point Keesing made. I agree with this point, but would like to take it further: We need to consider all our qualitative methodologies as different types of ethnographic encounter that will necessarily produce different kinds of discourse depending on the context.

For instance, in my cultural study of supermarket tabloids (Bird 1992a), I approached readers only after a thorough immersion in the papers themselves. For two years, I read them and spent time interviewing writers and editors, with the goal of learning how these texts came to be, and what were their standard conventions and subject matter. Ideally I would have liked to observe writers, but as with much research, I had to live with the constraints on access placed by those in power. My audience research was then in two steps. Following Ang (1985), I first solicited letters from readers, leaving the topic more open-ended than Ang had done, in that I included a request that readers write about "anything else they care to share." Many of the 116 letters, notably those from women, were personal, revealing documents, in which women explored questions not only about tabloids but about aspects of their own lives and their sense of identity. These letters helped me establish a sense of the people as far more than "tabloid readers" (clearly only one facet of their identity). Later, when I conducted the interviewing, I had learned the right words to use, and had a wealth of questions, many of which might not have occurred to me without their letters.

Many women clearly felt comfortable with the opportunity for self-analysis the letter writing offered. Men, on the other hand, wrote brief, informational letters with almost no personal or emotional content. Some asked for something definite to respond to, such as a survey or formal set of questions. As I analyzed the letters, I believe I reached some valid conclusions; their writing helped me more fully contextualize the tabloids as texts through which women explored important issues of self-worth, family, and caring, all of which had little to do with direct "response" to the specific texts (Bird 1992a).

However, I did not pay as much attention to the *process* of the letter-writing itself, and why it yielded such rich information for women, rather than men. In other words I neglected a consideration of the letter-writing as a form of "ethnographic encounter"/"communicative act" between myself and the people who wrote to me, and I feel sure that the very nature of the communication helped explain some of the differences I explored in my discussion of "gendered readings." For many men, letters are reserved for business matters, while women take care of social and family correspondence. Women are more likely to see letters as a personal form, and may be more comfortable in expressing themselves in a letter. At the same time, women may feel guilty about taking the time to write personal letters, or may believe that their opinions are not important. I was initially surprised at the number of women (never men) who thanked me for the opportunity to express themselves. Typical was this closing comment: "Thank you for wanting to know about me," or another: "Thank you for the opportunity to write to someone and sharing my thoughts and dreams of Life and the AfterLife—God Bless You." Perhaps I should not have been surprised; as McRobbie (1982) wrote some years ago, "Almost all feminist researchers have reported this sense of flattery on the part of the women subjects, as though so rarely in their lives have they ever been singled out for attention by anybody" (p. 56). I perceived in the letter-writing an echo of the "resistance" described by Radway with romance readers—that women enjoyed the opportunity to "indulge themselves" through writing a letter, an indulgence that could be justified in terms of "helping a researcher with her book."

I conclude, then, that soliciting personal letters worked better for women, because the idea of writing such documents made sense to them. In particular, it made sense to write to another woman, someone who had already defined herself as being in some ways like them, in that I was taking a serious interest in one of their pleasures—tabloids. Some, particularly the retired or home-employed, explained that they enjoyed writing letters to friends, "so why not to you too," as one put it. Others felt somewhat guilty, others were diffident about the value of their letters, but all seemed to feel comfortable with the process. Letters allow a certain intimacy without a possibly threatening personal encounter. They are a particular kind of communicative act which allows the writer to define the terms of the language, the length of time the communication takes and so on. In that sense, they offer an "ethnographic way of seeing," in which the participant is invited to define the terms of the encounter. The open-ended letter, diary, or other self-defined statement can be a very effective way to allow certain informants to do this. On the other hand, it may be precisely the wrong approach for others; undoubtedly, my invitation to participate in this way would have had little appeal for readers (men or women) who did not enjoy writing at all.

A second step of my research involved in-depth telephone interviews with both women and men. The letters had helped enormously in opening up the cultural context for my interaction with my respondents. During the interviews, I tried to conceive the interaction as a collaborative dialog, more conversation than interview. As Camitta (1990) writes: "The interview, as a social genre that is controlled by the interviewer, is a form of mastery over object, acquisition of knowledge through control of language. Conversation, on the other hand, is more collaborative, depending upon affiliation rather than upon separation in its structure" (p. 26). Once again, I found myself surprised by the richness of these conversations, especially with women. Telephone interviews receive short shrift in the literature on social science methodology. Typically, methods texts list the advantages of the "telephone survey," such as low cost, efficiency, convenience, and safety (eg. Babbie 1989, Chadwick et al. 1984, McTavish and Loether 2002). However, the disadvantages are said to be numerous. Chadwick et al., for example, argue that "Questions in telephone interviews must be brief and simple" (1984, 128). They elaborate: "It is difficult to conduct a telephone interview that lasts more than 15 to 20 minutes, and many respondents will tire before that point. In addition, telephone interviewing generally produces much less information than would be obtainable via personal interview" (p. 127).

They also decry telephone interviews as lacking in intimacy: "A further disadvantage of not confronting the respondent in person is not being able to establish rapport and a temporary sense of cooperation sufficient to minimize evasive or incomplete responses"(p. 127). Clearly, this language typifies classic social science assumptions—the idea of "confronting the respondent," "minimizing evasive responses" and so on does not suggest the cooperative enterprise of ethnographic research. I have no need to beat the dead horse of positivism further; what is more relevant is that feminist and qualitative researchers have, as far as I can tell, also bought into the idea that phone interviews are by definition a poor substitute for face-to-face contact. I know I did—I chose the phone because I wanted to talk to people around the country, and had no funding to support travel to do this.

Having embraced this "poor substitute," I was surprised and delighted to experience the rich, personal conversations I had with tabloid readers, especially women. Far from being necessarily short, several of the interviews were an hour or two, with women expressing pleasure in the experience, and even calling me back or writing a follow-up letter. As I thought about this, I began to see that this made sense when we see the phone interview not as related, but inferior to other kinds of interviews. Rather, we can see it as related to, and fitting naturally into women's oral culture. As Hopper (1992) puts it, in normal situations, "telephone speaking's rich communicative ecology is surprisingly like face-to-face speaking" (p. 10). "Phone surveys" are not like normal speaking at all. They are usually assumed to be calls made "cold," catching people unprepared and asking them to answer structured questions about topics about which they may care little. They are perhaps the quintessential example of what Hopper calls "caller hegemony"—that is, they use the telephone as a "summons" that "increases the power differential in the caller's favor" (p. 58).

More ethnographically-based phone interviews do not have to be that way, and can be set up to duplicate a normal social call as closely as possible. My phone interviews were the result of an initial letter from a reader, followed by a reply from me, with an invitation to choose a suitable time for a call. In other words, we went through a brief introductory period, where a tentative relationship was established, then continued on the phone. Rakow (1992) documents how for many women, the telephone is actually very personal, allowing confidences and secrets to be communicated in a uniquely intimate way. "Visiting" on the phone "is about the domestic, private sphere and takes places within it. It creates and maintains the relationships upon which families and the community are built" (p. 37). Essentially what I had done was set up a phone "visit" that was received as comfortable, unthreatening, and pleasurable. Had I been a man, I believe the response from women might have been very different, with the call perceived more distinctly as "business."

And in addition to different gender codes, as researchers we also often face different codes associated with class or ethnicity. Again, I found that the use of the phone interview shed light on this issue. Gal (1991) writes of the postmodernist critiques of ethnography, in particular the claim that ethnographies should be collaborative texts produced by researcher and researched: "What has received too little attention in all these critiques is the unavoidably power-charged verbal encounter in which anthropologists and native speakers, with different interests, goals, and deeply unequal positions, meet and attempt to talk" (p. 190). Gal may be talking here primarily about a Third World encounter in which the anthropologist is almost analogous to colonizer, but her point is relevant in other contexts. Most of the women I spoke with in my research were working class, with relatively low incomes (only one was an ethnic minority). I, although female, am white, upper middle class, relatively affluent, and British. In a face-to-face interview, that difference may have been much clearer, and could have stood in the way of the rapport we developed. Scholars advocating ethnographic interviews tend to suggest that these should be carried out on the interviewee's home ground, rejecting the traditional "experimental" model of bringing people to the researcher (see, for example, Agar 1996). Traditional social scientists will argue that interviewing on the informant's turf can be useful because it allows the researcher to study the surroundings, decor and so on. This may indeed seem useful to the researcher, but where does it leave the interviewee, whose most personal milieu is being studied?

Many years ago, Goffman (1959) wrote about "impression management," a task we all go about in our "presentation of self in everyday life." How we present ourselves to others is important to us all, and how we present ourselves *in our own homes* is perhaps the most important of all. Maybe I am shallow, but when I am expecting a visit from someone I do not know well, or from someone who has more power than me, I become anxious about how I look, how I act, how clean my house is and so on. Likewise, when I have interviewed people in their homes, I know they have often done a great deal to provide a "good impression." I think sometimes we forget the weight that the words "college professor" may carry. Those who themselves have social power are less likely to be intimidated, and more likely to have offices where they can meet you. Those with less social power are the ones who will be most affected by the "power-charged verbal encounter" of the ethnographic interview.

The phone interview, on the other hand, removes much of the power charge. Sarch (1993) describes how women sometimes found the phone liberating in dealing with the unequal power relations involved in dating. They felt at an advantage, "braver," or "more in control" when able to speak from the comfort of their homes. The very fact that a potential date (or a researcher) is *not* scrutinizing one's taste in clothes or furniture may be quite liberating. I conducted several of my interviews sprawled on a bed in sweatpants and T-shirt, and the women I spoke to usually sounded equally relaxed. As with the relative anonymity of receiving a letter, the researcher can find herself in the role of the "stranger on the train"—the faceless voice who is not and will not be part of everyday life, and thus to whom much can be revealed.

Like the letters, my phone interviews with men were briefer and much less intimate, seeming more businesslike. In my research, I concluded that men had a different approach to tabloids, treating them more as "information" and in a more detached way. I still believe this is true, and it is supported by other reception research (e.g., Morley 1986). Yet I do believe most men would find it hard to develop what might pass for appropriately close rapport with an unknown female researcher. They may also be more defensive about discussing an interest in tabloids, which are generally seen as a female genre. Neither the long phone call nor the personal letter are forms that fit well into male communication styles, perhaps especially for working-class men. Once again, I am left wondering how much my own gender, my choice of methods, and perhaps other cultural realities, played into the conclusions about gender reached in my study.

I am not suggesting some kind of blanket advocacy of phone interviews or letters for women. There could be many situations where they would not be effective. Businesswomen who use the phone all day at work may not relish the thought of a long phone interview. Those who write a lot at work may see writing letters as a chore, while those who have literacy problems might view a request to write as a threat or embarrassment. Some men may be very comfortable writing letters, or chatting intimately on the phone. All methodological approaches are characterized by constraints; the key is to be flexible, tailor one's "ethnographic encounter" to a particular situation, and be aware of the possible impact of methodological choices. Anthropologists are increasingly recognizing the importance of this basic methodological point, whether working in their own or other cultures. Like Keesing, half a world away in Melanesia, Goulet (1994) saw the nature of the ethnographic encounter as a key problematic. Discussing many anthropologists' misunderstanding of Arctic Dene (Athabascan) culture, Goulet concluded this was in large part due to their inability to go beyond traditional ethnographic methods, especially the formal interview. Dene people believe learning comes from personal experience, and do not perceive it possible to explain something to someone in "objective" terms. Thus, "To seek through interviews what Dene expect the investigator to learn through personal experience and observation will simply not do. Interviews are a poor means of investigation among the Dene" (1994, 119). Instead, Goulet developed a "narrative ethnography," creating ethnographic tales through mutual experience with the Dene.

And this more exotic example underlines the main point I want to make–that we need to be both creative and reflective about the methodological choices we make. For instance, we must think through more thoroughly the issues of ethnicity, and the differentials these produce in the research process. Ethnicity was one factor in my decision to use group interviews in a project on ethnic stereotypes in a popular TV program (Bird 1996). While van Zoonen is somewhat skeptical about group interviews, suggesting that they may produce a misleading homogeneity, I found them especially useful in discussing the program with Native Americans, whose communicative style is more collaborative and less competitive than Anglos, and who are culturally attuned to group discussions. On the other hand, mixed male/female Native American groups turned out to be somewhat less successful than single-sex groups, apparently because of a tendency for females to defer to the opinions of older men. Later, when I again used a group approach in my later study with Native Americans (chapter 4), I avoided mixed-sex groups.

So we cannot limit ourselves to one methodological approach. Field ethnographers do not carry out *only* interviews or *only* observation; they use any and all methods that are appropriate, including, one should stress, quantitative methods. Morley and Silverstone explain, "Ethnography is a multi-faceted process in which the requirements of detail and richness, rigor and systematicity, have to be carefully balanced, and where there is no single methodological procedure" (1991, 160). An essential element in that multi-faceted process is understanding which methods fit the cultural experiences of the people we study.

In addition, we should carefully consider the ideological work our methodologies are performing. Postmodern reflexivity often repositions the researcher at the center of the ethnographic endeavor, carried to its logical conclusion in "autoethnography" (see e.g., Behar 1993, 1996;

Ellis & Bochner 1996). While acknowledging the importance of this rethinking, I still see the "other's" voice as fundamentally more important in true ethnographic research. As Gal comments, " . . . it is not enough to insist, as the postmodernist critics do, that the ethnographic encounter and the genres that emerge from it are jointly produced . . . (This) omits the several levels of unequal power and privilege that characterize the ethnographic encounter and which also determine who is able to talk and what it is possible or strategic to say" (1991, 191). We need to work at developing any and all ethnographically-inspired methods to try to ensure that the voices of our research participants have at least equal time.

This goal inspired me to develop the "researcher-absent" methods used in chapters 2 and 4, in which I was interested in exploring quite diffuse aspects of media interaction. While most audience work on news has begun with a genre, or a specific text (Morley's *Nationwide* study being perhaps the ur-text here), I was more interested in exploring the idea of what news (in this case scandalous news) means to people, and how its stories "enter into life," as Nightingale (1996) had put it. These are questions that do not lend themselves easily to direct questioning, yet are also difficult to observe except in random fashion. We have no doubt all observed or taken part in social conversations about people or events in the news, but short of following workmates around with a tape recorder, we have few ways to capture the elusive nature of this interaction. With this in mind, I developed the technique of putting together a videotape of various types of news programming, and having people watch it with a friend or relative–someone with whom they already had a relationship. The participants then taped their own conversation, using some very general topics as guidelines, and I followed up with phone conversations with the participants. I believe the result, although inevitably guided by my interests, evoked a more naturalistic understanding of how people do indeed use news stories as frames to discuss cultural and personal questions. Seiter (1999) asserts that "often, conversation about the media is used as a pretence to talk about interpersonal relationships, longings, and desires, or 'taboo' topics . . . " (p. 2–3). I would argue that "pretence" is the wrong interpretation; rather, people work through these questions using the media as a framing and organizing device for their experience, just as other societies might employ myths or traditional gossip structures to do the same thing. This particular methodology, I believe, allowed me to grasp and demonstrate this rather elusive, cultural quality in a way that other approaches might not.

Thus, alongside current enthusiasm for self-reflexivity and au-toethnography, it is well worth thinking about how the *absence* of the researcher may actually allow us to capture a closer approximation of the natural context. This same impulse was behind the study described in chapter 4. In this study, I moved away entirely from direct interaction with any specific text. Instead, I was interested in exploring whether stereotypes that had been presumed to exist (through textual analysis and some direct audience studies) actually seem to be operationalized in people's lives. The "media creation exercise" asked a great deal of the participants; they were required to spend a considerable amount of time, with relatively little guidance, creating an imaginary television program. I hoped to find out what kinds of organizing principles they used, and how their cultural identity as White or Native American played into their relationship with media conventions. Essentially this methodology was an attempt to explore how any media text "holds within its structure signs of the history and the culture which produced it. The body of the viewer, too, is a cultural site which, in very different ways, reproduces signs of a personal history of engagement with that (same) history and culture" (Nightingale 1996, 125). While it is up to the reader to judge the success or otherwise of this method, I have been pleased to hear from two fellow-scholars who tried out this approach after hearing me present the work. Neither used it as a research method as such; both employed it in a classroom setting as a way to encourage students to articulate unacknowledged cultural scripts about race and gender, without confronting them with questions about "attitudes" and "stereotypes."

In contrast, I found that a hands-on, participant-observer approach was ideally suited to investigate the nature of a self-defined Internet fan "community"–and the more I was involved, the more my understanding of the group developed. Participating in fan writing activities, and developing virtual friendships with individual list members over a considerable period of time was invaluable not only in developing a genuinely ethnographic sense of the community, but also in breaking down the barriers between "me" and "them." Mere interviews, whether virtual or face to face, could not have been as effective. This kind of long-term familiarization also proved important as I wrestled with the question of how audience study might shed light on the thorny issue of aesthetic judgments in popular media (chapter 5).

Finally, chapter 6 forsakes direct audience interaction altogether, as I develop a cultural history of a particular "text," or narrative, drawing on the insights of folklore and mythological studies, and suggesting the value of exploring the "shared agendas" (Speer 1992) of folklore and popular communication studies. This case study cried out, not for narrow investigation of how specific people "read" this text, but for a much broader exploration of the dynamic interplay between media and oral communication, and between overarching cultural archetypes and specific historical moments. For in truth, our everyday lives are an inter-connected mesh of communicative acts, which link us with media, with each other, and with our historical traditions. Carey (1989) called for a rediscovery of everyday forms, "making large claims for small matters: Studying particular rituals, poems, plays, conversations, songs, theories, and myths, and gingerly reaching out to the full relations within a culture or a total way of life." (p. 64). My chosen methodology for this study was planned to be this kind of cultural analysis; my goal was to capture an "ethnographic way of seeing" the folklore/news nexus, even though no direct response study was involved.

CONCLUSION: THE VALUE OF MEDIA ETHNOGRAPHIES

In thinking seriously about the methodologies employed in my various case studies, I am not suggesting that my way was the only way. But I do suggest that if different methods had been used, my studies would have been different. And if similar studies had been carried out by other researchers, they also might have been different. Might I have succeeded in gaining richer insight on male tabloid readers with another method? Quite possibly. Could I have learned something about response to scandalous TV news by sitting individuals in front of a TV and then asking them questions? Probably. The methods I chose would probably have been less effective in a culture in the early stages of media saturation (see for example Kottak 1990), since they tend to assume a kind of deeply in-grained, naturalized media literacy that is not the norm across cultures, but which in the West is taken for granted and often unproblematized. I chose these approaches because they seemed to hold potential to shed light on my basic anthropological question: What is it really like to live in a media-saturated culture?

I hope the coming chapters will offer insight into that question—because with all the reflection about how methods play into conclusions, my core belief is that it *is* possible to learn about culture through ethnographically-inspired methods. As Agar writes: "Ethnography no longer claims to describe a reality accessible by anyone using the right methods, independent of the historical or cultural context of the act of describing." However, he continues, "On the other hand, there is no justification for . . . complete relativism . . . There is a human group out there who lived in a world before the ethnographer appeared and who will continue to do so after he or she leaves" (1986, 19).

Some ethnographic narratives are better, more true, than others, and while we might have lost the certainty of our right to "speak for the other," we should not abandon the goal of finding better ways to understand the dynamics of culture through the eyes of others, as well as our own. For me, writing about others' experiences is infinitely more interesting and rewarding than writing about myself, even as I recognize that my research is inevitably refracted through my personal identity. In some senses, ethnographic audience research is indeed impossible; we cannot reduce the lives we study to a text. "There may be a correspondence between a life as lived, as life as experienced, and a life as told, but the anthropologist should never assume the correspondence nor fail to make the distinction" (Bruner 1984, 7). The confidence once felt by ethnographers is gone, and will never return. Turner's observation that we are now "*groping* towards an experiential or 'inside view' " (1985, 223, emphasis added) is closer to the truth of contemporary ethnographic experience than the confident conclusions of ethnographers who aspired to grasp the "native's point or view." It sounds rather messy—which is of course the beauty of it, as ethnographers leave themselves open to the possibility of the unexpected, and the excitement of sudden insight.

Taking methods seriously is one way to assert how important it is to keep working to find better, more nuanced ways to represent and understand the qualities of our mediated culture. In my final chapter, I return to a concluding discussion of issues of power and constraint, and the future of media ethnography. Meanwhile, in laying out my own forays into media ethnography in the next few chapters, I hope to demonstrate how these studies, carried out at different times and with different people, nonetheless come together to suggest an agenda both for studying "the audience" as real people, and for looking "beyond the audience" toward a richer ethnographic understanding of life in a mediated world.

2

MEDIA SCANDAL MEETS EVERYDAY LIFE

INTRODUCTION

No matter how often media scholars and columnists scold them for it, people *like* scandal. Scandal sells newspapers and tabloids, keeps people in front of their televisions, and provides endless opportunities for conversation. This is not a new phenomenon; the penny press, early tabloids, and "jazz journalism" thrived on scandal. Of all the news stories current in 1989, the top two in terms of audience recognition were the conviction of Oliver North and the trial of the Rev. Jim Bakker, while the major story with least recognition was catastrophic health care coverage, which surely touches the lives of many more of the news-watching public (Price and Czilli 1996). Two of the highest-rated news interviews ever, at least before the O.J. Simpson trial, were Barbara Walters' conversation with Donna Rice in 1987, and Diane Sawyer's with Marla Maples in 1990 (Mellencamp 1992). [1] The Top Ten stories in audience interest identified by the Pew Research Center for the People and the Press are consistently personality-rather than issue-driven, although not necessarily scandals. In 1998, the Clinton-Lewinsky scandal became perhaps the biggest U.S. news story ever (at least until the qualitatively much different September 11 catastrophe), and it also marked the arrival of the Internet as a site both for new information and for popular dissection of the scandal. Three weeks into the scandal, Lewinsky drew "as many people as Princess Diana and the Super Bowl to major media websites like the Washington Post, AP News and CNN. New Monica sites had their traffic zoom from zero to more than 15,000 daily hits within hours, while existing sites reported traffic jumping from 10,000 hits to 100,000 hits a day. And the "Drudge

Report," which was averaging 55,000 hits a day, skyrocketed to more than 300,000 daily hits" (Koplowitz 1998).

The emphasis on the personal, the sensational and the dramatic has always been a central element of news. Street literature, ballads, and oral gossip and rumor have all contributed to the development of news as we know it (Bird 1992a; Bird and Dardenne 1988; Stephens 1988). Critics have been pontificating about the salacious excesses of newspapers for generations; even mainstream news has always been torn between what practitioners see as a duty to inform, and their need to entertain and engage their audience.

Existing surveys confirm the preference for human interest stories, while at the same time describing many people's feelings of guilt about this preference (see, for example, Price and Czilli 1996). In the most recent of its periodic assessments of news interests, The Pew Research Center concluded that "In 1997, Americans were happier with their own lives, felt more religious conviction and were less attentive to the news than at any time in recent years" (1998, 1.) From 1986–96, 25 percent of respondents closely followed the news stories tracked by Pew, while only 19 percent did in 1997. The death of Princess Diana was the only story that attracted the close attention of a majority of people surveyed, while "for the second consecutive year, not one domestic policy story made the annual list of top 10 news stories" (p. 1). Parker and Deane (1997) write that although only small percentages of people claim to follow stories about scandal and entertainment, the data suggest "that the public *knows more* about these types of stories than it does about virtually any other category of news. On average the public answered 60 percent of the questions dealing with scandal, entertainment and crime correctly" (p. 6.), a percentage far higher than any other category.

Critics often seem baffled by this. How can people spend so much time thinking about such obviously "trivial" topics? Part of that bafflement comes from a difference in the way journalists/critics and audiences define news and how it is used. Journalism critics tend to define news in terms of how effective the texts of news stories are at conveying information about the world to readers and viewers. They assume that readers consume news in order to learn facts about the world around them and be informed. In that respect, they follow what Carey (1989) calls a "transmission" view of communication. Audience definitions also include this "informing" function, although I believe the cultural pressure to be informed is felt less and less today. But everyday definitions of news focus more on how these stories are inserted in people's daily lives, growing and becoming the subjects of speculation and discussion.

As Dahlgren writes: "Audiences can take the stories and 'run' with them, in many directions" (1992, 15). Much of the news that readers and viewers are exposed to is either ignored or forgotten almost immediately; from the audience perspective, relevant news consists of stories that take on a life of their own outside the immediate context of the newspaper or television broadcast.

Stories that do take on life tend to be dramatic and personal—these are the stories that people actually remember, as I shall show. This is not necessarily a bad thing; the rise of talk shows, tabloid news, and other forums has allowed concerns to be raised that otherwise would not be. Showing the personal side of public events is probably the most effective way to make people understand the impact of those events. And people want personal stories because they are memorable. Anthropologists have shown that the kind of stories people remember are chronological narratives, with a clear structure, a moral point, and vivid imagery. Traditional inverted pyramid news stories, with the standard "who, what, where" format, are the most difficult to remember.

Yet while the popular appeal of human interest news is widely acknowledged, there has been little serious attempt to understand that appeal from the point of view of the audience. Critics tend to agree that the audience has a role in determining the direction in which journalism is going (with this direction often decried as a process of "tabloidization" (Sparks and Tulloch 2000), but limited empirical work has been done on how audiences actually view news, how they define it, and what they do with it. In fact in most cases, with the obvious exception of such studies as Morley (1980) and Jensen (1990), analyses of news make assumptions about the audience from the texts themselves. In this chapter, I apply data from two small, empirical studies to offer some observations about how people interpret and use scandals in their everyday lives, as a step toward understanding the pleasures they derive from such disreputable stories. Bringing the discussion up to date, I explore how the Internet has facilitated this kind of daily talk on a wider scale, perhaps helping to move scandal even further toward the top of the news agenda.

Langer (1992) efficiently documents what he calls the pervasive "lament" among media critics about the state of contemporary news, especially television news: "In its unease, the impulse of the lament is to act, to 'clean up' television news in order to get rid of the unworthy elements, relegating them to the dustbin of journalistic history" (p. 114). The "unworthy elements" that are commonly decried are those that are defined as "sensational"—personality-driven stories focusing on people who flout society's norms, whether these are celebrities or "ordinary people" thrust into the limelight by remarkable events. The supermarket tabloids, often popularly called "scandal sheets," have come to epitomize the worst of the unworthy.

What is often lacking in the "lament" about news is any real understanding of *why* scandalous or sensational news is appealing. Some critics have applied psychological explanations—defining the audience as "sensation seekers" who need increasing doses of exposure (Zuckerman 1984) or as "morbidly curious" (Haskins 1984). These explanations tend to have the effect of neuroticizing the audience, suggesting that there is something sick or abnormal about being attracted to unwholesome news. Or if the audience is considered at all, it is often to condemn them as lacking in taste and judgment, as Langer discusses. More recent, and in keeping with trends in media reception scholarship, cultural studies has discovered the "active" audience for scandalous or sensational news. Thus Fiske (1992) and Glynn (1990) celebrate the tabloid style in print and television, arguing that audiences may epitomize De Certeau's "textual poachers" (1984). The "excess" in this kind of news can be "transgressive" and "calls up sceptical reading competencies that are equivalent of the social competencies by which the people control the immediate conditions of their everyday lives" (Fiske, 1992, 54).

STUDYING THE RECEPTION OF SCANDAL

As I have written elsewhere (Bird 1992a), I am not convinced that the skeptical, carnivalesque reading of tabloid style is actually typical of most consumers of this kind of news, although it is definitely one way to read these texts. Rather, I am more inclined to see audiences as active, selective readers, who approach all kinds of news from an unstated perspective that essentially asks: "What can I get from this information, or this story? How does it apply to my life, and why should I pay attention?" In conceptualizing the reception study that built on my earlier work on tabloids, my goal was to explore responses to news of all kinds, with no special emphasis on scandal, in an attempt to throw some light on the role that news stories play in everyday life. What kinds of stories do people find memorable? What do they do with them? This chapter draws on qualitative data from this project and from my tabloid research, aiming to place "scandalous" news into a larger cultural context.

My goal was to elicit data that at least approximated everyday experience. In the Spring of 1993, I prepared a videotape made up of excerpts from the then-popular tabloid TV show, *A Current Affair,* the "reality-based" show *Unsolved Mysteries,* and an episode of ABC's *News with Peter Jennings.* Copies were lent to a small sample of people (22), who watched the tapes in their homes with a family member or friend, and then discussed the tapes, recording the conversation with a small audio tape recorder left with them. Some guiding questions were included, asking, for example, which stories they found most memorable and why. Later, I interviewed the same people by phone, asking them a range of similar questions, such as what the idea of "news" meant to them; which kind of news stories they paid attention to, and so on. Through all of this, I was trying to get a sense of how news fits into people's lives on an ongoing basis, drawing both on their observed and somewhat more naturalistic conversations, and on their more self-reflective opinions that emerged from interviews. Recently, I have supplemented this approach with monitoring of news message boards, in which people discuss current issues. Although these issues range across the political spectrum, it is striking how scandalous and personal news attracts the largest interest; as I write this in early October 2002, the Yahoo news message board shows that over the previous week, over 14,000 messages were posted about the disgrace of New Jersey Senator Robert Toracelli, as compared with, for example, just over 1,000 discussing a story about rising numbers of people being without health insurance, as unemployment and recession hurt the economy.

NEWS AND STORYTELLING

The first step to understanding the role of "scandal" is to recognize that this kind of news is invariably cast as "story," rather than the terse, inverted pyramid "news" style. I have discussed elsewhere the difference between these two, and reviewed some of the literature regarding audience preference for "story" news (Bird and Dardenne 1988). Media scandals may begin with a short news item—Hugh Grant has been arrested following an "encounter" with a prostitute—but the full-blown media scandal only develops with the follow-up stories. How could this clean-cut actor do such a thing; does he have a dark side; how does his girlfriend feel; if you can't even believe in a nice guy like this, who can you trust, and so on. Not all potential media scandals develop like this; but there do seem to be certain characteristics of the media scandal that need to be present, a point to which I will return. For now, I will focus on the story dimension of news, because there is a larger set of narratives to which scandals belong—all media scandals are stories, but not all stories are scandals. I believe this is the key to the enduring appeal of scandal—that it is *one* type of narrative that helps people structure their view of what the world is and how it should be.

But *are* stories actually important to people? Do people care how they get their news, or do they just want to be informed? Conventional journalistic wisdom suggests they do—all journalistic textbooks encourage student writers to "find the story," to humanize it and so on. Price and Czilli (1996) point to a substantial literature that confirms this, as well as providing their own data demonstrating the memorability of human interest stories. Participants in my 1993 study consistently reported that the news items they remembered from the tape and those they wished to talk about were the *Current Affair* and *Unsolved Mysteries* ones. Even those people who said they despised such shows preferred to talk about them. By contrast, most respondents found it hard to remember any of the *World News* stories, even though they had just sat through them. Respondents aged over 35 were far more likely to say the more traditional *World News* approach was better, more accurate, and what news was about. For these viewers, being informed was a value in itself, yet even they were not moved to talk with enthusiasm about hard news. Compare, for example, these two exchanges between a middle-aged (40s), married couple. First, they discuss the *World News* segment.

> *Wife:* Oh yes, there was something about the economy.
> *Husband:* They were discussing the economics, where Clinton says he's going to do something about it, and the Republicans say, well, don't fix it if it ain't broke. And more or less to that effect...
> *Wife:* Well the Republicans kept saying things were getting better, and the Democrats kept saying they weren't...
> *Husband:* Yeah.
> *Wife:* And then the growth for the last three months of the quarter were 3.8, which is a pretty good...
> *Husband:* Which is a pretty good growth, yeah.
> *Wife:* Growth. Umm...

Later, the same couple talks about the segment on *Unsolved Mysteries*, which involved a story about a purported UFO landing. The difference in their tone is clear:

> *Wife:* I think the show about the UFO was interesting in that there are so many unanswered questions.
> *Husband:* Yes, why did the guy get burned? Was he burned because of the radiation that was in the ground, apparently there before?
> *Wife:* But why in circles? Or buttons, as he called them himself...
> *Husband:* Where he got burned. Yeah, I don't know.
> *Wife:* And I was surprised...

Husband: Yeah, and why did his shirt catch on fire when that ship took off? How does that, how did that occur, it looked like he was a decent distance away from it.

Wife: Oh, he was running away from it, though. They didn't show that they found the burned shirt, that's, I wondered if that was anywhere around...

Husband: And why did the compass go bonkers? There's lots of questions...

The couple went on to discuss the story animatedly for several more minutes, returning to it later. In contrast, participants generally found it difficult to discuss hard news stories at all, except in terms of their personal interests. This applied especially to those who expressed a clear disdain for news as "boring," the word used most by younger respondents. Males in their twenties were more negative about *World News Tonight* than any other group, reacting strongly to its lack of dramatic appeal. Even the perceived importance of a story did not change that, as this exchange between a college student and his friend illustrates:

Respondent: The, uh, all the ABC *World News Tonight,* that sucked.

Friend: Yeah, they were lame.

Respondent: Waiting for those verdicts in the LA riots. That was kinda interesting, I mean, 'cause that was such a big national deal. But the story wasn't that great.

Friend: Yeah, not that interesting.

It was only when a story actually hit a personal chord that these young men conversed about the program at all. For example, a segment on the Clinton "don't ask, don't tell" policy on gay people in the military, which tended to be ignored or dismissed by older respondents and women, made for a few minutes of conversation:

Respondent: Do you think it's bad that they're gonna maybe be in there?

Friend: I mean it's just like, I mean there's gay guys...you know, military and everyone's close and your showering together and stuff.

Respondent: I'd hate it.

Friend: I mean, yeah, you never know if some gay guy's looking at you.

Respondent: Pick up that soap for me man.

Friend: Yeah, you're in the shower, you ain't gonna be pickin' up no soap.

This pattern continued throughout the taped conversations. Older people expressed a concern for being informed, while younger people often aggressively rejected the very notion, but both groups became engaged when stories spoke to their personal interests or convictions. In follow-up interviews, this preference for the personal story came over just as strongly. Thus a woman in her 50s says:

> Well, I'm interested in people. I just like people and I like to know about them ... That little girl that fell down the well in Texas ... The boy that had his arms got caught in the threshing machine ... I'm very interested to see how things come out on those kind of cases.

A male college student explains his preference for talk shows:

> This is the actual people talking about their problems, and you get questions from a bunch of people, instead of, you know, three or four ideas from producers or whatever, so you get all these people asking questions from different angles.

A 39-year-old woman says that she likes stories because, "Life without emotion is life without feeling, life without forethought, and life without good results. You've got to feel in order to react." She continues, "There's a lot of reasons why things happen, a lot of reasons why people do the things they do, and if a person understands the story behind it, they can have a better idea and a better awareness to avoid being in that situation themselves." These comments point to the way that news stories are applied to a person's life, an issue to which I will return.

A younger woman, a student in her early twenties who enjoyed *A Current Affair,* explained what kind of stories catch her attention:

> "Usually things about people being in trouble, as far as, you know, like getting involved in a murder or like a big sex triangle, or like, I read the *Wall Street Journal* lately, and like people making a bad choice, and then all of a sudden they're involved in a fraud ... and how their whole life now is a shambles ... "

She continued, explaining the importance of the story:

> "With that storytelling, they seem to, they start from the beginning, like if they do have a person there who is in this big news event or whatever, they'll like start from them when they were real little, and them growing up, and them getting married, and you find out stuff from their background, and they just make the whole or-deal sound like something that could happen to anybody ... and

on *Current Affair,* they give every little news item, they give it like a little name ... That kind of makes it seem more of a good story."

Her comments point to one of the key elements of the satisfying story—that it is a coherent narrative that has a point to it. She notes that *A Current Affair* titles its stories, as do news magazine shows; the title functions to sum up the story and package it. Media scandals tend to gradually cohere into "a story" as they develop, so that after a while, the public recognizes the central narrative associated with "Jesse Jackson's illegitimate child" or "the Amy Fisher case."

A college student in his early twenties also explains the way certain news shows are able to direct the unordered flow of a narrative into a "story," offering detailed accounts of features from favorite shows like *Rescue 911,* many of which aired days or even weeks before, while being unable to recall stories from the previous night's evening news. As he commented,

> "I mean people want to see what happened and how it happened, and when you're watching the regular news, I mean, they just tell you what happened, and not how it happened. I mean, they might show you a couple scenes, for example, like Waco, but they won't recap on what David Koresh was thinking, or what happened to him ... I mean I don't even know if they found the guy!"

A woman in her thirties referred quite specifically to the value of "storytelling" in news, pointing to her preference for talk shows and magazine programs like *20/20:*

> "It's different than the basic news that you receive, you know, at 5 or 6 o'clock. Storytelling would be just people living their lives, and what's happening in their lives, and how they're bettering their lives, or how they're screwing up their lives, or whatever, and I guess that I'm just more interested in that type of thing, and what's happening more on a personal level ... I guess I just find it more interesting to listen about something someone personally has achieved, or whatever."

Others, while dismissing shows like *A Current Affair* as being "trashy" or "sleazy," admire magazines like *60 Minutes* because they are "more dramatic" or "more personal."

Thus while many people feel an obligation to be "well-informed" and listen to "important" news, their emotions and their attention are caught by dramatic, exciting stories. These do not have to be scandals—but any story with special human appeal. Tabloids are full of heartwarming stories about ordinary people who become heroes for a day, or about beloved celebrities who overcome drug problems or obesity. News magazines and "tabloid" TV shows mix scandal and "upper" stories comfortably. Sagas like the rescue of Baby Jessica McClure from a well (Pew's top story of 1997) are as big ratings boosters as the scandal of Amy Fisher.

Thus to put scandal in context, we need to consider it as one element in a mesh of stories through which people construct and interpret their lives. Indeed, as I see it, a barrier that stands in the way of our understanding of media scandals is that too often we tend to separate out types of media narratives both from the array of other types of narrative, and more broadly from the culture of which they are part. We look at the text of the news story or the TV show, analyzing what it says, and what it might mean for the audience. Instead, we need to think of these stories as emerging out of the culture, and as then sparking a broader set of interrogative narratives and discussions among the people. [2] For as Johnson writes, in defining cultural studies, a full understanding of any text must take into account the complexity of the relationship between the three components of producer, text, and audience. He argues that this relationship should be seen as circular rather than linear, in that producers incorporate readers in their production of texts, texts in turn may have an impact on readers, whose response then feeds back into the text (Johnson 1983).

I have argued elsewhere that media and oral storytelling are comparable, though not identical, communication processes, during which narratives are constructed from familiar themes that repeat themselves over time. And if audience members are seen as active in helping to shape the way popular media are created, they become much more comparable with folk "audiences." Even in an oral culture, not every individual is a storyteller or an active, performing bearer of all traditions. Rather, the role of people in many contexts is to respond to the storyteller, helping her or him shape future versions of the tale. The popular media audience can play an analogous role.

In certain media genres, the kinship with folk traditions is absolutely clear. The most sensational of the supermarket tabloids, for example, draw deliberately on folklore, and the beliefs and concerns they know their readers already have. According to a former tabloid reporter, "When looking for ideas for stories, it's good to look at fears, and it's good to look at real desires. That's why a lot of people win lotteries in the stories, and why people get buried alive all the time . . . " (Hogshire 1992). Tabloid writers rely quite heavily on tips from their readers in developing stories, some of which can grow into long-running narratives fed by reader interest and participation, such as the "Elvis is Alive" narrative that has extended over years. The more "respectable" tabloids, which focus on celebrity gossip, borrow less directly from folk narratives, but nevertheless clearly shape their stories according to the standard themes that work for audiences, such that the specific stories change, but the overarching narratives recur. The lives of individual stars become molded to the established repertoire of celebrity sagas, retelling the folk tales that preach the dangers of hubris and the lesson that money and power cannot buy happiness. While more mainstream news practitioners like to maintain a distance between themselves and tabloid styles, clearly their human interest stories adhere to many of the same narrative conventions (see essays in Carey 1988).

Media scholars, then, have learned from folklorists and anthropologists that culture is participatory, rather than coercive. In a classic article, Bascom (1954) argued that folklore serves to educate audiences in the values of a culture, validate norms, and also allow an outlet for fantasy and wish fulfillment. I would argue that news, especially "story" news does the same thing, involving and drawing in its readers and viewers. Different types of stories perform any one or more of the same kinds of functions as folk stories do.

SCANDAL AND THE AUDIENCE

Indeed, in many ways the notion of "scandal" is more firmly embedded in the oral, interpersonal dimension of our lives, rather than the media dimension (although these are closely intertwined). The media play the role of the storyteller or town crier, but the scandal gains its momentum from the audiences. Gluckman (1963), in a classic discussion of gossip and scandal, describes the role of individuals known as *simidors* (leaders), in Haitian villages who lead "scandal" songs sung at working bees, in which the indiscretions or failures of people in the community are lampooned. People fear their power; as one Haitian put it, "The *simidor* is a journalist, and every *simidor* is a Judas!" (p. 308). It is not so much the song itself that is feared, but rather the way the story now becomes the public property of the village, and source for endless speculation. Some songs tend to die, while the subjects of others may fuel gossip and discussion for weeks. Likewise, many news stories could potentially spark larger "media scandals" but not all do. To become a real scandal, the media accounts must spark the imagination of the public.

Price and Czilli suggest that one reason people remembered stories like Oliver North and Jim Bakker is that these stories simply received a great deal of coverage. Similar arguments were heard during the apparently-endless Clinton-Lewinsky scandal—that the media somehow forced the story on the public, who had little option but to respond. This explanation comes from a linear perspective on media communication: The media produce a text, and the audience hears and responds. If instead we see the relationship between media and audience as more circular, then we would note that increased coverage certainly would have raised awareness of the stories, and of course the public would not even have heard the story in the first place without the media. However, the increased coverage was also a direct result of the high level of interest and speculation among the audience that were being generated.

In fact, as Mellencamp (1992) begins to suggest, the initial media coverage may actually emerge from gossip and speculation among the community, as in the case of Jim and Tammy Faye Bakker: "The gossip about their private lives, initiated by Jim's illicit sex with Jessica Hahn and Tammy Faye's drug addiction, fueled the story and the legal machinery..." (p. 222). Media stories then led to more gossip, spinoffs into other scandal forums and so on: "The Bakkers were tabloid fare which made it to the nightly news; a personal catastrophe was transformed into a media scandal, created and uncovered by gossip...." (p. 222). Mellencamp then chronicles the rise and fall of the Bakker media scandal in television, tabloids, newspapers and news magazines, until "The end had come for Jim and Tammy Faye Bakker—that is, until the made-for-TV movie" (p. 229).

Dramatizing Morality

And why do some stories turn into major scandals, while others do not? Compelling stories must have a *point*—they must mean something. Folklorists and anthropologists who study oral narrative assume this without question, even if the narrative concerned is primarily told for entertainment rather than for moral edification. Of the many thousands of stories that are told in any culture, only some catch the people's imagination, and enter the tradition. Since the oral process is selective, scholars assume that those stories with a long "shelf-life" are significant in some larger sense—they speak to the moral values, fears, or fantasies of the people. Media scandals, or any media story that has staying power, should be explored in the same way, for the values and boundaries they are expressing.

The scandal story, above all, interrogates morality, as Gluckman suggests. Other anthropologists also document the role of gossip and scandal in drawing moral boundaries. Antoun (1968), for example, describes the complex system of satire and gossip that punishes unacceptable behavior in a Jordanian Arab village, while Gilmore (1987) shows how conformity is enforced in a small town in Andalusia, Spain, by an informal mechanism of gossip and backbiting that aggressively attacks people who violate the status quo. Martin-Barbero points to the way media texts have long been incorporated into oral culture in Latin America, with people commenting on the stories, and applying social values to them: "It is a listening marked with applause and whistles, sighs and laughter, a reading whose rhythm is not established by the text but by the group. What was read was not an end in itself but the beginning of a mutual acknowledgement of meaning and an awakening of collective memories that might set in motion a conversation. Thus, reading might end up redoing the text in function of the context, in a sense, *rewriting* the text in order to talk about what the group is living" (Martin-Barbero, 1993, p. 120). In particular, Martin-Barbero underlines the need to comprehend the oral roots of the tradition of melodrama. "The stubborn persistence of the melodrama genre long after the conditions of its genesis have disappeared and its capacity to adapt to different technological formats cannot be explained simply in terms of commercial or ideological manipulations." Popular media narratives, he argues, are essentially melodramatic, emphasizing morality and excess: "Everything must be extravagantly stated, from the staging which exaggerates the audio and visual contrasts to the dramatic structure which openly exploits the bathos of quick and sentimental emotional reactions. . . .Cultured people might consider all this degrading, but it nevertheless represents a victory over repression, a form of resistance against a particular 'economy' of order, saving and polite restraint" (p. 119).

Scandal stories, indeed, are not "polite," and as they gather momentum, they become increasingly unrestrained. Connell (1992) points to the importance of the scandal story in drawing moral boundaries between what is acceptable and what is not, with celebrities acting as larger than life, melodramatic personifications of correct or illicit behavior. He discusses a British scandal in which Jimmy Tarbuck, a popular comedian with a wholesome family image, was exposed as an adulterer, and suggests that the public views celebrities as a privileged class. "It is status which grants rewards, but it also grants responsibilities . . . Tarbuck's affair has been disclosed because he has chosen to ignore these responsibilities" (1992, p. 81).

Regular tabloid readers (interviewed in my earlier study) clearly agreed. A 60-year-old woman argues that celebrities should not expect privacy, and deserve scandalous exposure if they behave immorally: "If you're a celebrity you have to expect this . . . listen, once you're a celebrity you don't have anything to yourself, you belong to the public . . . and I think a lot of them, they're a little out of line as far as the ways they live... I think they should live a life that will be respectable. So I think if you want privacy, go lock yourself up somewhere." Others concurred: "I think once you're in the public eye, in any capacity, whether you're a politician or a movie star or a TV star... You're fair game... they'd better be clean or they'd better stay out of it." Yet another woman talked about the then-marital problems of Prince Charles and Princess Diana: "They make too much of it, and it's sickening... we need something to look up to... they're not it, and they don't deserve what they have." Stories are clearly seen as comprising useful checks on public excess, much like the licensed gossips of oral cultures: "If people know they're going to be spread across the front page of the *National Enquirer,* I really do think it may not give them pause to think at the time, but it certainly gives them cause to think later..."

So the stories as they appear in the press, particularly in tabloids, and tabloid-style TV shows, make the point about morality clearly and often melodramatically. Yet these melodramatic tales also leave plenty of room for discussion and comment; while melodrama often places good and evil in stark contrast, the audience becomes involved in the speculation of how and why such shocking and exciting things came to happen. Actual "applause and whistles, sighs and laughter" may not accompany the watching of *Current Affair,* but the discussion afterwards may incorporate many of the same emotions. Media scandals, at least the ones that have the longest staying power, often are making an over-arching statement about right and wrong. Jimmy Tarbuck is an adulterer, and that is wrong, especially given his image. Jim Bakker and Jesse Jackson are licentious and immoral, and that is wrong, since they should stand for Christian morality. Oliver North is a liar... But wait, one hears a reader ask. Didn't Ollie lie for the very best of motives? Wasn't Bakker driven to adultery by his freakish wife's drug habit? And he did repent, didn't he?

We see this interrogation in online discussions of the Jesse Jackson scandal, which erupted in January 2001, with the *National Enquirer's* revelation that the Rev. Jackson had an illegitimate daughter with a former campaign aide, to whom he had been paying unofficial child support for some time. This became a popular topic on the *Enquirer's* own news message board, "Planet Tabloid," as one of many topics posted in the Politics section (despite the fact that the *Enquirer* does not cover hard news, the topics on the message board run the same gamut as those on Yahoo, CNN, and other "news" boards). By following the discussion, we can see how the story is used, not as a "text" with a clear meaning, but as an opportunity to interrogate issues from morality, to religion, to race. An important point that becomes clear in this thread is that, especially in forums that support anonymity, people may abandon the code of civility maintained by those who know each other and interact in an environment such as the workplace—for instance expressing overt racism. Nevertheless, the points raised and positions taken are similar to the discussions on my tapes; people do not evaluate news stories in isolation, but incorporate them into their already-established worldviews. The point of contributing seems to be less about trying to persuade others to one's point of view, but rather to clarify one's own position, and perhaps enlist and identify allies. The Jackson thread opened, and most messages were posted over a three-day period in January 2001, when the story broke, followed by a second flurry in February, following continuing coverage. For space reasons, I have included only those messages from the January thread, and have edited some for length. This sequence of exchanges is quite long, but I believe a detailed transcription is useful in demonstrating the full flavor of the Internet news "conversations" that are now so much a part of news in everyday life. I identify posters by an initial, showing that while most contributors post only once, several engage in back and forth posts. Original spelling is retained:

A: Hey Jesse . . . This is typical of all liberal democrats. Just a bunch of hypocrates and babies. . . . Sorry liberals, this will be the most horrifying 8 years . . . because Bush will be elected again . . . so just get over it. And to Jesse Jackson . . . you out of all people deserve it for all your witch hunts, i hope you burn in hell . . .

B: Very sad situation. I mean, Rev. Jackson is a Democrat, just like the rest of us. By us, I mean the popular vote. Lets FOCUS on what the Republicans do wrong. There are many things.

C: Another false prophet succumbs to the temptation of Satan.

D: You LIBERALS RUN YOUR MOUTHS, THINK YOU DESERVE SPECIAL RIGHTS, AND THAT THE REPUBLICANS ARE TO BLAME, CAUSE YOUR MEN CAN'T SEEM THE KEEP THEM ZIPPERS UP. THEN YOU STUPID "I MEAN STUPID" WOMEN TELL THE WORLD IT'S OKAY IF HE CHEATS . . . THE DNC PARTY IS OVER . . . YOU WANT FAIRNESS LIBERALS, MOVE TO IRAQ.

E: I doubt very seriously that George will be in office 8 years. First of all, he didn't win this one, HE TOOK IT. Jesse Jackson is a minister true, but his ***k still rises, so whats your point?

F: Let's hope you're still around when George W. makes a complete ass of himself. And by the way which is the greater sin. Snorting coke and drunken driving or an adulterous affair?

G: At least Jesse admitted to the affair and will support the child. That goes a very long way in my heart towards forgiving this man. He told his family before the story broke in the N.E. Hilary didn't know about Monica until she'd heard it on the news. I forgive Jesse Jackson and will continue to pray for all of the people involved.

H: you people have all missed the mark . . . . . if those damn idiots (ie. Jesse Jackson, Al Sharpton and Kweisi Mfume to name a few . . .) would stop dipping their pens in the company inkwell . . . we wouldn't have all these problems. Where did these self-proclaimed heros go to school anyway ? Have any of them even graduated high school or GED?

I: I luv the so called black leaders. They take pride in making girls half their age pregnant. Jesse Jackson. Willi Brown. Qweisi MFume . . . . . This is great. They are producing what they can and are good at. The human race and animals have been doing the same thing since nature started ??? Dont blame them. They dont know what to do with technology, science, medicine, producing goods and services etc . . . They are doing what they do best! Throwing a ball through a hole.

J: Thank you Jesse for sending the image of a Black man whirling back a century or so. It is not enough that Blacks have to spend everyday of their lives overcoming stereotypes . . . . When you put yourself out in the public as much as you do, you have to realize that you become a role model for others . . . I use to see him as a strong Black male who loved and respected his wife, family and himself. Now is just another male who does not know how to keep his pants zipped.

K: I never held that high of an opinion of Rev. Jackson. However, I believe many people in our country did . . . They put their trust in a leader who basically turned out to be a hypocrite. I can't imagine how I'd feel if one of the political leaders I admire . . . turns out to be someone completely different . . .

L: All GREAT men are Fallible too. The fact that Jesse Jackson has an illegitimate child does NOT negate all the GOOD he has done in this country . . . Even Martin Luther King was an adulterer . . . Don't fall into the white man's trickery and judge Jesse by his enemies exposure of this very private matter. Notice how this was just exposed right before George bush takes office. The republicans were terrified of Jesse . . . Their trick did'nt work with most Black Americans . . . Jesse has devoted his life to fighting racial injustice . . . I have never heard Jesse acting like he was a saint.

G: AMEN! Why is everyone holding this up as some major sin when our own president did much worse while in the white house? I can forgive Jesse Jackson. I can not forgive Bill Clinton. Look all of the tax dollars he spent trying to hush all of action up as well as lying to the American people? . . . Jesse is a far better man than Bill Clinton anyday!

I: What about the $40,000 Jesse gave her for moving expenses? Moving money or hush money?

L: Rev. Jackson didn't set the example. Our White forefathers did. They bastardized us African Americans when they impregnated slave women . . . I guess all of America has forgotten that illegitimacy is as much a part of our heritage as anything else. I'm not lending righteousness to adultery, because it is in violation of one of the Ten Commandments . . .

M: Perhaps our downfall as human beings is that we foolishly look for that which is only God's, in other human beings.

I: yes. the evil white man took his dick out and made him have sex with that woman. It's all a conspiracy. I'm sick of people's indiscretions being blamed on someone else. Leave race and political views out of this. Who is to blame? Nobody but Mr Jackson and that woman!!

H: Jesse is a racist and so are many Extreme Leftist so called Black Leaders. There are some good black leaders on the right with Solid values like Colin Powell, Dr. Thomas Soul, Larry Elder, Justice Thomas etc . . . Unfortunately most blacks are brain washed to follow destructive leaders on the left like Jesse . . .

N: I . . . am stepping back to consider that just maybe Ms. Stanford was single, intelligent, and facing her biological clock . . . and wanted to have a child, and discussed this with her friends and family, and also with Rev. Jackson (she was 37 at the time), and rather than go to a sperm bank or get any other donor, Rev. Jackson by mutual choice became the donor . . . Let's say it was done in a doctor's office or hospital or sperm bank, as artificial insemination . . . He makes no secret of this, the National Enquirer gets the gist of the story, twists a few facts, and the media run with it . . . If this is the way it really is, it is totally consistent for Rev. Jackson to keep his role private, and admit that he fathered a child and has taken financial and emotional responsibility. It was the media that used the term "affair" and other descriptions that could turn out to be distortions . . . It could be.

G: I look to God . . . I am a sinner, so why should I look to another human (sinner) as a role model for me? . . . No matter what, if God can forgive all for our sins, why can't we forgive one another?

I: And people like you fail to face reality and drink from the river of denial. FACT is 1) Jackson BONKED a broad without the consent . . . of his legitimate . . . wife. Makes him an A**HOLE IN THE FIRST DEGREE. 2) Jackson paraded his Ho around Washington DC while relentlessy spitting fire of damnation onto like minded people. Don't play with fire Jesse, cuz you WILL get BURNED. 3) If he has nothing to hide then let there be an audit of the books from The RainBow Coalition to make sure his financial support squandered on Mistress Ho and baby girl wasn't money donated from mostly impoverished members . . .

O: The picture on the front page of the National Enq is great. There is Jesse next to his girl grinning from ear to ear. Ms Jackson should roll up a thousand or so copies of the magazine and beat the____out of Jesse

Speculation

A scandal that is relatively long-lived must enter the public conversation. In doing so, it tends to follow a fairly standard pattern. First, there is *speculation:* Yes, something is wrong, but why did it happen? If a rock star known for dissolute ways was caught with a prostitute, as Hugh Grant was, there would little public interest, since it is what we might expect. When John Entwhistle, aging bass player for the Who, died in 2002 of cocaine-induced heart failure, reportedly in the arms of a prostitute, the circumstances caused barely a shrug, as commentary focused on his contribution to a rock legacy. As Greenblatt (2002) writes, "Nothing says rock star like Vegas hotel room, a high-class hooker, and loads of blow." On the other hand, the Hugh Grant story was huge because it was both excessive and incongruous. The incongruity surrounded the lack of fit between the public image of the clean-cut, gentlemanly British movie star, who maintains a stable relationship with a well-known model, and the man who turns to a prostitute for sex. And this is compounded by excess; he doesn't call a high class escort service for "regular" sex, but is caught in the back of his car with a "cheap" hooker fellating him. The Lorena Bobbitt story had the same elements, in that we see the incongruity of a young, apparently mild-mannered married woman not merely hitting back at or even shooting a possibly abusive husband, but actually cutting off his penis. Not stopping there, she throws the offending member in the street—melodramatic excess *par excellence*. These stories invite speculation and moral judgment, as well as the laughter, jeers, and ribald commentary of the melodrama audience, especially if they include "trademark" details like a severed penis, a stained dress, or a cigar. As it grows, the saga takes on the quality of a dialog among members of the public, and between them and the media. The scandal story, then, is not clear and closed, but "open"—allowing for many competing versions and interpretations, as the Jesse Jackson thread shows clearly.

A cardinal rule of journalism has long been to learn how to put a human face on current events. As Tomlinson (1997) writes, "Personalization can be read here not as trivialisation but as achieving greater imaginative proximity to the lifeworld of the audience" (1997, 77). While journalists and media critics wring their hands because the public "needs" to be informed and is apparently perversely resisting this need, people themselves say, "Why do I need to know this; what difference does it make to *my* life?"

Furthermore, elite definitions of news and popular definitions are often at odds, in that news that would be dismissed as salacious gossip by critics may be perceived as useful information by audiences. The very questioning and speculation invited by scandal may help people discuss and deal with issues of morality, law and order, and so on, in their daily lives. For all the distaste many people had for the Clinton-Lewinsky saga, it became a catalyst for widespread examination of relevant social issues, including extensive parent-child conversations (Pew Center for Research on People and the Press 1998.) An earlier case that came to the fore in my study was the 1992–93 Amy Fisher/Joey Buttafuoco scandal, which chronicled the trial and imprisonment of 16-year-old Fisher for the attempted murder of her middle-aged lover's wife. Amy Fisher drove critics to despair. Pointing out that indexed clippings of the story topped 100,000 by 1996, Krajicek writes: "For most of the 1990s Buttafuoco's every belch appeared as news in hundreds of newspapers and on scores of television news programs in the United States, Canada, and abroad . . . he was covered relentlessly, beyond all reason" (1998, 15). Each of the three network television stations produced a movie about the case, and a survey showed that by January 1993, 40 percent of Americans had seen at least one of them. (Parker and Deane 1997). Buttafuoco was given massive coverage again in December 1995 when he violated his parole. "Since 1992 the mass media had been attached to Buttafuoco like barnacles. If I had chosen to do so, I could have had informed conversations about Buttafuoco with friends in California, relatives in Nebraska and Florida, and colleagues in Toronto" (Krajicek, 137). Of course, that was exactly the point. Several participants in my study raised the case as an example of distasteful sensationalism; even though it was not included on the study tapes, they were almost irresistibly drawn to discuss it. A snippet of conversation between a mother and her teenage daughter points to the conscious way in which people participated in the story, which was open to a range of interpretations:

> *Daughter:* As many people as watched the Super Bowl watched the Amy Fisher story on all three networks. It was on all the talk shows, it was on all the news headlines. *Current Affair, Inside Edition.* And that was how America lived for four weeks, or however long.
>
> *Mother:* And didn't we like it because we could be the judge?
>
> *Daughter:* Well, yeah.
>
> *Mother:* Didn't we like it because we were being the judge of that? We could point our fingers whichever way we wanted it. We could make it Joey's fault, we could make it Amy's fault, we could make it her folks' fault. We could make it anybody's fault we wanted. We could be the jury.

As people speculate, they tend to look for answers from within their own experience, and their level of interest is often connected to how closely they can relate the scandalous events to their own lives. What would I do if this happened to me? How can I prevent this happening to someone I know? From an audience point of view, the best stories are those that leave room for speculation, for debate, and for a degree of audience "participation."

In particular, people of different ages and backgrounds applied the story differently in their lives. Thus, a 67-year-old woman said:

> "Oh, I think it's rather creepy. Let me tell you, I'm retired now, but I was a social worker. And I guess that I worked with a lot of people . . . People . . . a little bit lower in morals, and so forth . . . and I guess at this point I'm kind of hoping he gets his, too . . . I think we all know things like that go on, and again you can look at it from several standpoints as the woman is blamed, although certainly she was wrong, and I just have an idea that he probably egged her on . . . so I think he's equally guilty. And I think she was, oh, you know, just a young gal who thought everything would be OK if she just got rid of the wife. Naive. Even though she was well-versed in sexual matters."

A second woman, age 39, stresses different elements in the story:

> "It was just an amazingly huge story, and it was this young girl, older man, and I think what made it kind of interesting was the wife being so strong behind her husband, and I found that kind of fascinating myself, because everything just seemed to stack against him. But yet, this woman just really steadfastly, you know, stands behind him. And Amy steadfastly stands behind her story, too, they all do. And it's, I guess it's whatever you want to believe, and everybody takes it their own way . . . And to think that a young girl would actually, you know, shoot another woman in a love affair, I guess . . . "

A 21-year-old woman perhaps reflects her closer sense of identification with the teenage Amy Fisher:

> It's a thing I could never imagine happening to me, being involved in a murder, or having this big love triangle or something . . . I mean, how did Amy get into that situation, that she had to shoot that sleazy guy's wife, I mean you'd have to be desperate, and he must have had some kind of hold on her—he was just slime, don't you think? I mean really, his wife didn't deserve that . . . If I ever thought my life was bad, at least I'm not this person . . .

Finally, a woman in her forties saw the story as a vehicle for speculation that led into a consideration of how people allow this to happen, and what sort of circumstances throw ordinary people into these extraordinary events:

> You know we watched a couple of 'em here, my kids and I. Because, I don't know why we did, I think because we had read so much about it, and then they had the different sides, and it was sort of like, we watched it and then we decided who we really thought was the guilty person, and who gave us the facts and who didn't, and . . . but it's garbage, I mean, really, it's garbage.

When asked why this story was so interesting, she continued:

> Just maybe human nature, I guess . . . like when I hear about stuff like that, what I always ask myself is how, you know, how could they get to that point . . . show me why this person got to this point . . . it's, you know, you cannot relate at all to what this person has done or why they've done it, and so you want this explanation . . . And maybe that's because you're afraid that your next door neighbor could end up [like] that, you know . . .

Participation and Personalization

Scandals, like other kinds of news stories, are clearly integrated into this woman's life, allowing her to interrogate boundaries, questions of motivation, and issues that seem much more relevant to her life than, say, economic news. She describes how she discussed with her family the 1993 "Spur Posse" scandal, in which the media focused on a group of white youths from "respectable" homes, who awarded each other points for having sex with as many girls as possible:

> Oh, it made me sick ... I've got high school aged kids, and then a daughter that's in college, and especially the high school age, they talk about things like that in their classes ... She reads everything she can, and that really interested them, so it's kind of fun to sit down and watch something with them ... They may be able to relate that to certain guys or girls ... it's like, tell me this isn't happening in your school.

This personalizing of "story" news is crucial to understanding people's enjoyment: An important aspect of this process of personalization is that it is participatory. As we struggle to make sense of a story, we involve others in the negotiation of meaning. Thus a 21-year-old man said, "When you watch by yourself, a lot of times you have ideas that you have unsolved because you can't converse with other individuals." A 37-year-old woman agreed: "I feel that in order to really be a good conversationalist, you can't be self-centered, and I want to hear everyone's opinion about what's going on in the news. There's something in their view that I can use, and hopefully there's something in my view that can contribute to making theirs better." She considers *Unsolved Mysteries* to be important news because "it helps others in the community feel a part of the news world ... The community or the listeners get to contribute to the story and make the news effective and be part of the results."

A pair of short excerpts from the taped conversations between participants give a flavor of how people explore the meaning of scandalous stories together. The first is a snippet of conversation between a 29-year-old male office manager, and his male friend, discussing a *Current Affair* story on the videotape, about a woman who persuaded a teenage boy to murder her husband:

> *Participant:* I mean how can you, in a normal fashion, walk up to someone and even hit him? Let alone shoot him four times in the head?
>
> *Friend:* I guess that's maybe its value to any one of us knowing ... I mean ... here's this woman that appears to

the neighbors to be great with all the kids... everybody's pal, then she doesn't get along with her husband... I mean...

Participant: Yeah, there's a lot of non-related things to bring up, so you could suspect everyone in the world, and that's like, wow, you know, our neighbor, she sure seems nice—could she kill her husband, and what would make her do that?

A husband and wife in their early forties discuss the same story:

Wife: We're just noticing how, the progression of how the seed was planted in the kid's head, and by this woman, and how it just grew, and she kept feeding and encouraging it... But the saddest part is that this kid, he's kind of backed in a corner by these three girls saying hey, let's do it. And the next thing you know, he shoots this guy...

Husband: Yeah, one of the things that struck me as they were interviewing the kid in prison was that he didn't show any remorse or sorrow other than the fact he couldn't have children. He never once said how bad he felt about killing the guy...

Wife: Yeah. My own, like, my biggest thought was, you know, what kind of family life did he come from, what would put a person, a teenager in this kind of position... to be so bland of caring and emotion?

The stories that become larger scandals are often not simple, in that they do not have closure, and there could be more than one answer. The Oliver North story, for example, was so big not because it was a simple morality tale, but because people disagreed on what the story really was, and so many questions remained unanswered. In large part, it played out as a "misunderstood hero" story (Andersen, 1992). But not everyone thought North was a hero; some saw him as a liar and a cheat. Speculation ran rife about why he did what he did; which powerful politician knew what, and when; where the glamorous Fawn Hall fit in; and so on.

Likewise, the Gary Hart scandal was a story that invited participation from the public. The spin-off from the story was not just a simple lesson that politicians should not have affairs. After all, many had done so in the past, and their reputation was unsullied. Hart's scandal was not just about morality, but about whether personal morality or lack of it was an issue in a political campaign. Gripsrud (1992) writes that morality scandals around politicians are clearly understandable: "for one thing, it is not without public interest if the minister of defence is a paranoid junkie" (p. 92). That view certainly seems accepted now—yet not so long ago, people were not so sure that personal moral failings were relevant to political achievement. John F. Kennedy was actually protected by the press corps, who knew all about his adulterous affairs, but did not consider them relevant to his abilities as a president (Gans 1979). Hart's story involved extensive interrogation, both in the media and among the public, about the importance of marital fidelity as a qualification for political office—the "character issue." In some ways, the story marked a watershed, in both public opinion and media coverage of politicians' personal lives.

A few years later, Sen. Bob Packwood became the focus of another watershed scandal, as the media *and* the public rejected the notion that sexual harassment was normal and acceptable. Yet Packwood's behavior was not that different from the way other politicians had behaved in the past, without censure, and certainly not everyone agreed that the moral lessons of the story were cut-and-dried. In my study, Packwood's case was mentioned in a brief *World News Tonight* item, and while most participants passed over the story as unmemorable, a 22-year-old man discussed what he saw as the implications of cases like these:

> I mean right now, the way things are going, men are, white males are becoming a minority more than anything with saying things and pretty soon we're going to be doing the same thing women are doing. I mean it's getting to the point where you have to sit down and state what actually is prejudice and what is slurring and so forth.

The Pleasure of the Story

The point here is that scandal stories, like other stories, bring changing mores into sharp focus through media narratives and the popular discussion that takes off from those narratives, whether in homes, workplaces, or on the Oprah Winfrey show. Media scandals help set the agenda for discussion, but they do not exist as some definable text separate from the wider cultural conversation. Thus the biggest media scandals are open texts—they draw some lines about morality, but they do not answer all the questions. Often the most popular scandalous stories are those where there is debate—who is worse, Amy Fisher or Joey? Who is the victim, Charles or Diana? Was Lorena Bobbitt an avenging angel or a crazy devil? Indeed, with Clinton-Lewinsky, the biggest scandal story in recent memory, the issues raised in the popular debate were essentially also those that consumed political pundits, pointing again to the artificiality of drawing clear boundaries between producers, text, and audience, and between media analysts ("us?") and media audiences ("them?"). For instance, "feminists commenting on the ... affair remain deeply divided on the question of whether Lewinsky was a victim or a perpetrator; whether her experience proves that women can use sex to get power or that women will always be used for sex by powerful men. The enduring disagreement about her status renders Lewinsky fascinating as a subject of feminist analysis" (Williams 2001, 97). And these are only a few of the many open questions that swirled around the affair both in the media and among the public.

Langer analyzes the appeal of "victim" stories—the human interest tales about ordinary people who are caught up in accidents, crimes and so on, rather than scandals. He argues that these stories provide three kinds of pleasure. The first is seeing "the social and psychological mess when the orderly arrangements of everyday life collapse" (p.126). Coupled with this is the "pleasure of uncertainty" (p. 127), in which people enjoy the vicarious awareness of random danger in the world. Finally there is "narrative pleasure," as sensational news "can more readily foreground its story-like constructedness by positioning its reader/viewer into a kind of mutually confirming declaration with the response: that was a good one!" (p. 127). Scandal stories evoke the same kinds of pleasure, as audiences both revel in and are repelled by the social mess that is created when the famous transgress moral norms, like Hugh Grant or Jim Bakker, or when an "ordinary" person like Lorena Bobbitt does something so bizarre.

This kind of pleasure is analogous to the way people listen to personal tales, urban legends and so on. Ellis (1989) describes the questioning and playing with belief that often accompanies the telling of such tales, as the story demands "that the teller and listener take a stand on the legend: 'Yes, this sort of thing could happen; 'No, it couldn't'; 'Well, maybe it could'. . . a legend is a narrative that challenges accepted definitions of the real world and leaves itself suspended, relying for closure on each individual's response" (p. 34). A 69-year-old male tabloid reader describes much the same kind of response: "We do hand 'em round each other and then we get to talk about the stories and then we'll say, did you read about that thing in there! And of course it makes a little conversation and then we'll say, Oh, you don't believe that, do you?!"

The enthusiasm with which audiences participate points to another dimension of the news story in general and the scandal story in particular. People say they like human interest stories because they identify with the people in the stories, whose plights engage their emotions. This does seem to be true, and this emotional engagement continues in certain "big," human-interest stories, such as the Jessica McClure rescue, the story of the recovery of Oklahoma City from a terrorist bomb, or perhaps most of all, with the deluge of human stories about heroes, villains, and victims after the September 11, 2001, terrorist attacks (Bird 2002).

Distancing

This level of engagement clearly entails a personalization of the story, and a sense of empathy with the subjects. However, many scandal stories move past a sense of personal empathy, creating a distance between the audience and the story. This seems to happen at a point at which a general consensus has been reached about what the "end" or the "meaning" of the story might be—what the moral lessons actually were. Langer raises this question: "We might speculate then, that at the ideological level, the recognition of story-ness may act, not to engage the viewer/reader in the victim story's premises and potential outcomes, but to produce distantiation: these are real people, here is misadventure, but after all, it's only a story."(pp. 127–28). This is an important insight, that has relevance to the media scandal, and to its reception as melodrama. Melodrama, as we have noted, is about excess—the "I can't believe it!" dimension of life— just as are urban legends and jokes. Once something becomes seen as an over-the-top melodrama, the people caught up in it begin to seem less like real human beings, and more like cartoons or symbols. Scandalous celebrities, already larger than life through their celebrity status, become even more the property of the public. People feel free to speculate about the most personal aspects of their lives, using them as props to work out their own moral codes.

This distancing becomes most apparent in the way media scandals evolve so easily into jokes. Even though the news media tend steadfastly to ignore the jokes that cluster around the stories, disdaining them as tasteless, these joke cycles become an integral part of the public discourse on major news events, whether disasters or scandals (see, for example, Goodwin 1989, Oring 1987). Almost everyone has heard jokes about Ted Kennedy and Chappaquiddick, Gary Hart, Tonya Harding/ Nancy Kerrigan, Michael Jackson, O. J. Simpson, and now, of course, Clinton-Lewinsky.[3] The existence of the jokes presupposes a wide knowledge of the particular story; the jokes then contribute to the firming up of the story in particular ways, confirming Gary Hart as a lying philanderer, Michael Jackson as a pathetic pedophile, O. J. Simpson as a man who got away with murder, and Bill Clinton as an incorrigible womanizer.

As Davies writes, "Jokes are ambiguous forms of discourse that are created in circumstances and around issues where there is a good deal of uncertainty" (1990, p. 8), just as media scandals surround events that provoke speculation. But as he also points out, as they develop, jokes tend to reflect the prevailing moral judgment of the public: "the sense of sudden vicarious superiority felt by those who devise, tell, or share a joke" (p. 7). There is an exhilarating satisfaction to be gained from transforming Hugh Grant from the envied guy who has it all to "the Englishman who went up Divine and came down his pants." With the widespread circulation of jokes, the scandal reaches a kind of resolution, with the subjects of the joke/scandal becoming caricatures with whom we have little sense of identification as people. Significantly, when people still feel pain about a major event, such as the Oklahoma City bombing or September 11, few if any jokes emerge.

CONCLUSION: THE PLEASURES OF SCANDAL

The term, "media scandal" is in some ways a misnomer. Certain news stories become true scandals only with the participation of the media audience and the public at large; people know about major scandals even if they have never read or seen a single news story about the event. Certainly the media are now the major tellers of scandalous tales, but the narratives cannot be sustained by the news media alone. The narratives must speak to issues or emotions that engage readers and viewers in speculation, fascination, and downright relish in the melodramatic excess of it all. Thus in spite of its frequent designation as a "scandal," the Clinton administration's Whitewater affair never really became one in the public perception. It lacked drama, human interest, a clear sense of any moral codes being broken, and any excitement whatsoever. Neither the media nor the general public had a clear idea what the "story" of Whitewater really was. There was no clear or compelling speculation and gossip, and there were few if any Whitewater jokes. The Clinton-Lewinsky affair, in contrast, was clearly a multi-dimensional morality play, interrogating issues of sex, power, gender roles, trust, and so on. Interestingly, clear victims and villains never quite solidified; all major players, from Clinton, his wife, Lewinsky, Kenneth Starr, and even Linda Tripp were seen by different people as villains. It has been speculated that one reason for the continuing support of Clinton was that the public never accepted Lewinsky as a victim (Sonner and Wilcox 1999). When the energy finally went out of the scandal, the "story" that appears to have gelled in public consciousness, expressed through jokes and casual comments, is one of Clinton as a flawed, weak womanizer, who frittered away his chance at true greatness, Lewinsky as a slightly pathetic "bimbo," and Starr as a self-righteous prig, and a reluctant acceptance among pundits of the "inevitability of scandal politics" (Quirk 2000).[4]

I have tried to place scandals in cultural context, seeing them as playing out moral dramas in extravagant terms. If news audiences generally prefer lively, dramatic, human interest stories over news about political and economic issues, this is not necessarily a terrible thing. As Dahlgren writes, news can and should be pleasurable, even though "the discourses of journalism cannot admit this" (1992, 16). He goes on to argue that "journalism ... often *does* foster ... feelings of collective belonging ... yet this is rarely recognized and even more seldom praised" (p. 17). The conversations that viewers and readers have about news stories serve to bind people together, and give them common topics of conversation in a world in which common ties are getting fewer and fewer. Scandal, and the delight we feel in it, has always been an integral part of news, and the many forms of communication that preceded "news" as we know it. The question now becomes: Is the enjoyment people derive from scandalous news something to be celebrated or decried? Fiske has been a leading exponent of the school that sees popular culture as liberatory (Fiske 1989a; 1989b). Contrasting the news of "the people" with that of "the power-bloc," Fiske argues that "popular taste ... is for information that contradicts that of the power-bloc: The interests of the people are served by arguing with the power-bloc, not by listening to it" (1992, 46–47). Thus tabloid news sources, such as *A Current Affair* or the *Weekly World News* can be viewed as offering criticism of the power-bloc, through the skeptical laughter that these shows produce, as authoritative news is inverted and parodied. Popular news, then, can function as social and political critique, unmasking and mocking the powers-that-be.

Fiske's analysis of popular reception of tabloid news has much to recommend it; he clearly recognizes the oral, participatory dimension of that reception. In addition, his view of popular news as carnivalesque and excessive meshes nicely with a reading of popular news as melodrama. However, his argument ignores the fact that tabloid news is not always received with skeptical laughter and disbelief, although it may be. Stories about scandalous government cover-ups of UFO landings are taken seriously by many people (Bird 1992a)—hardly a progressive or politically critical stance to take. More important to this discussion of scandal is that Fiske misses the point that melodrama and carnival, like much of oral culture, tend always toward the maintenance of the status quo, as do media scandals. The functions of folklore that Bascom and other folklorists describe all work toward the validation of norms, and the restating of accepted values. In addition, the anthropological literature on gossip and scandal clearly indicates their role in maintaining conformity. This is not to say that these norms remain forever unchanging; oral narratives, like ongoing media scandals, do interrogate norms, allowing for questioning and some gradual changes. As we have seen, the interrogation of morality that accompanied the Gary Hart scandal produced a shift toward a more rigorous standard of personal conduct among politicians, such that the entire landscape of American politics has now altered.

Nevertheless, one would be hard-pressed to conclude that highly popular news stories, whether scandals or not, have ever had any major subversive or liberatory outcome for "the people." If anything, they serve as distractions that help "the power-bloc" get on with the business of maintaining power. Fiske himself recognizes that whether or not popular news has any real political impact is "a difficult question" (1992, p. 54), arguing that "the pleasures of scepticism and parodic excess can be progressive," but "I do not wish to suggest that they are always or necessarily so" (p. 54). In fact, the only concrete example he offers for the progressive value of popular news is the TV talk show, using an example from *Donahue,* in which a disaffected Drug Enforcement agent sparks an audience discussion of the dangers of drugs, and the possible role of the government in exacerbating the drug problem.

While Fiske sees this kind of discussion as a hopeful, progressive sign, the trend in popular news seems rather to be in the direction of ever more "stories" that actually have little to do with political critique or resistance. As Campbell points out in his study of *60 Minutes,* the appeal of that show is that it tells formulaic, neat, dramatic stories that engage the viewer's interest (Campbell 1991). At the same time, it and shows like it inevitably reduce complex events to a tale of individual agonies and triumphs. A story about a homeless person becomes detached from the structural and institutional situation that put that person there, and becomes simply a tale of a personal setback. Radio and TV talk shows of the kind that Fiske admires for their liberatory potential, revolve around personal experience, which is what makes them so gripping. They do indeed allow many voices to be heard, and they certainly encourage excessive and melodramatic exchanges—but how does that translate into effective political critique or involvement?

My concern, then, is that "story" news, instead of being one element in news, is moving toward becoming the only element. Hallin (1992) makes a compelling, if slightly alarmist case that there is a growing "knowledge gap" between those who consume serious news, and those who like only gossip, scandal, and personality stories. If the latter group forsakes serious news altogether, he argues, this will have serious consequences for democracy.

My own work with audiences of tabloids and news audiences does suggest that there are some people who avoid hard news altogether. But it also suggests that most do not. People read the *National Enquirer* and *Time;* they read local newspapers and watch *Unsolved Mysteries.* But clearly, the stories that stay with them are those that engage emotion as well as intellect, and which provide material for rewarding interpersonal communication. Scandals certainly do that. Are they pushing out serious news to the point at which public discourse is threatened? At present, I believe the answer to that question is a qualified no. The public conversation does not revolve solely around issues of scandal and personality; people still discuss political issues and complex economic questions. Indeed, people are quite capable of separating juicy, fascinating scandal from other aspects of the political and economic arena. One of the most intriguing aspects of the Lewinsky scandal was the way Clinton's approval ratings on competence (not morality) remained high; in fact much of the public disapproval was reserved for what was viewed as hounding by the Starr investigations (Sonner and Wilcox 1999). However, the media do have a crucial role in setting the agenda for public discourse, and with the proliferation in personality-driven news, and a decline in newspaper readership (Hallin, 1992), the agenda is in danger of becoming increasingly trivial. In this chapter, I have tried to show how scandalous news is pleasurably useful, in that through these media morality tales, people come to terms with their own moral codes and values, as well as enjoying themselves immensely. All cultures need to do that. But if personal morality tales are *all* we are telling each other, this surely could have implications for the future of a functioning democracy, and I will return to a consideration of those implications in my final chapter.

3

PIECING A CYBER-QUILT
Media Fans in an Electronic Community

WHAT IS A FAN?

Most people, most of the time, are fairly casual media users. Surrounded by media on every side, we pick and choose the moments when we really pay attention, and genuinely become involved. As we have seen in chapter 2, those moments may happen when something in the news piques our interest—perhaps when a current news story seems to speak to our own experience. If we are members of a particular ethnic group we may have heightened awareness of media representations of our group; our political or religious persuasions may focus our attention on particular texts or images. In other words we constantly and largely unconsciously sift through endless media moments to determine which ones actually matter to us. And many of us find that at some moment or other, we have become "fans" of a particular TV program, book series, band, genre, or favorite performer. Even within the domain of "fandom," there are all kinds of commitment levels, ranging from, say, a habit of regularly watching a particular TV program, but not caring too much if we miss it, to a level of devotion that sees us spending many of our waking hours interacting with the object and/or with other like-minded people. In understanding the way media are integrated into contemporary culture, we need to explore the phenomenon of fandom, and the pleasures and commitments it entails. Being a fan can be an important cultural marker—a way we can signal our preferences and demonstrate connections with others. There's that ubiquitous question: What is your favorite movie, TV show, band, football team? When we learn someone is a fan, we feel we have learned something about his character; when we discover a friend is a fan of something we hate, we might stop to wonder if we really have so much in common with her. And if someone is *really* enthused about something, especially connected with popular media, we might feel there's something not quite healthy about this.

So what does it mean to be a "fan?" With its roots in the word "fanatic," the term drags heavy baggage with it, both in popular and academic discourse. Newspaper stories about fans usually include at least one expert chiding them to "get a life," such as an article quoting a sociologist who believes television is "substituting and sucking up the real world" (Owen 1999, 1). Jenson (1992) describes the way "fandom" is routinely pathologized: Fans are "obsessed with their objects, in love with celebrity figures, willing to die for their team. Fandom involves an ascription of excess, and emotional display" (p. 20). Fans are disreputable and even possibly dangerous. Ethnographic analyses of media "fan cultures," such as Bacon-Smith's (1992) and Jenkins's (1992) studies of *Star Trek* culture, have gone some way to providing a more thoughtful discussion of fan activities, as they describe the often creative practices of "Trekkies," who interact at conventions, through mail and through shared products, such as fan fiction. Nevertheless, the perception remains that fans are to be pitied or avoided; if their shared enthusiasm provides mutual enjoyment and a sense of connection, surely that is only because their lives are so otherwise empty?

FAN CULTURE IN THE ELECTRONIC ERA

Until recently, most academic and popular accounts of media fans have focused on highly active and visible groups such as *Star Trek* enthusiasts, with their outlandish costumes and conventions. The last few years, however, have seen the realization that electronic communication, specifically the Internet, has opened up an enormous range of new possibilities for media fan activities, many of which are much less visible, and thus more private (Baym 1997). People who would never dress as Klingons for news cameras may nevertheless feel a close connection to a television show or movie series, and have found ways to share that connection with like-minded fans all over the world.

For several years, I have observed and participated in one such electronically-connected television fan group, working toward an anthropological understanding of the role of such groups in contemporary culture. I have been exploring several related issues, all of which have become the focus of widespread debate as research on electronic communication has burgeoned. In this chapter, I ask several related questions.

Can the study shed some light on what it means to be a "fan," and could fan activities actually be a much more positive and constructive part of life than is usually perceived? Has electronic communication enabled fans to change and intensify the fan experience in creative ways? Or conversely, has it allowed fans to slide even further into what might be defined as obsessive behavior? Following an episode, for example, members may receive up to 200 commentary messages during a single weekend. To keep up, members have to maintain a sustained interest in the show and all the intertextual phenomena clustered around it, although they are not obviously and overtly participating in a fan activity, such as a conference or gathering. The whole experience thus becomes more internalized and "virtual." At the same time, the electronic medium enables fans to produce a wider range of creative activities, including imaginative role-playing games, fan fiction, Web pages and so on. The ease of communication allowed by e-mail makes collaborative activities much easier and faster than more traditional modes of communication. How is this intense "virtual" experience both like and unlike more conventional face-to-face interaction, and what implications does that have for understanding social interaction in our contemporary "wired" world, at a broader level than simply the study of media fans?

Related to this, I explore whether an e-mail list or other kind of electronic group can actually function as a "community" for its members. Will a long-term study of such a group help us better understand what constitutes "community" in the electronic age? What are the limitations and the potential of the electronic community—a question that goes beyond the more narrow issue of fan communities in particular?

And I also touch on the issue of "virtual ethnography," a topic I have explored in more detail elsewhere (Bird and Barber 2002). How is doing ethnography in a "virtual" community different from doing it in a "real" community? Can a consideration of this difference help us understand something about the nature of contemporary culture? The substantial methodological and ethical issues around Internet research are only recently becoming the subject of serious debate (Jones 1998), but my own experience tells me that ethnography in the virtual environment is both similar to and different from in the "real world," and presents its own unique set of challenges.

THE LIST

The group in question is an e-mail discussion list, DQMW-L, created for fans of the now-canceled CBS TV program, *Dr. Quinn, Medicine Woman*. Like much ethnographic research, my choice of the DQMW-L list was somewhat serendipitous. The project spun off from earlier research I was doing on representations of American Indians in the show; I became involved with the group when a member found and posted an article I had written to the list, and invited me to respond to a discussion of the article (Bird 1996). This was an exciting and rather worrying opportunity for me. In the rarified world of academe, we rarely find out what people actually think of our work, or indeed if anyone even reads it at all. How many academics have the chance to watch their work being discussed in a forum of people who are not fellow-scholars looking for validation for their own work, but who have an entirely different set of loyalties and interests? In any event, the discussion was enlightening and humbling; list members called me on things that were very justified; they gratifyingly said that my work helped them see their show from a different perspective, or that it opened their eyes to the way an American Indian might see it; or they castigated me for political correctness. And I learned that this was a lively, erudite, and interesting group of women; a few months of list membership convinced me of the interesting possibilities for ethnographic research.

From that point, I became a participant-observer on the list; although I mainly "lurked," I also posted from time to time, and participated occasionally in the fantasy "stones trips" that will be discussed below. Once I had decided to study the group's dynamics, I announced my intention on the list, and invited comment. I received no negative response to this, and subsequently I occasionally reminded the members of what I was doing, and posted drafts of my study conclusions every now and then. Members were invited to respond to these drafts, which many did. Several list members have established roles as "experts" in particular areas, such as women's history, the frontier, or specific characters, and they are expected and even asked to comment in their "specialties." I believe my role as an established member, regarded as knowledgeable in Native American issues, helped ease any concerns that members may have felt about "being studied."

Dr. Quinn, Medicine Woman premiered in Fall 1992 to overwhelming derision from critics, who described it in terms ranging from "treacle" to "drivel . . . its every nuance calculated out of lowest-common-denominator concerns" (See O'Connor 1993; Zoglin 1993). It was a frontier drama about a pioneering woman doctor (Jane Seymour) in Colorado Springs, set in the 1870s, and exploring her relationship with the various townspeople, her adopted children, and the man she eventually married and had a daughter with, "mountain man" Byron Sully (Joe Lando). In spite of the apparent critical message that only an idiot could enjoy the program, it became the most successful new drama series of the year, and developed a huge fan following. It ran for six seasons, and its cancellation in 1998 surprised many in the business, and devastated its fans. Fans then spearheaded a campaign to reverse CBS's sudden cancellation; while unsuccessful in reviving the series, they did help ensure that two additional *Dr. Quinn* TV movies were made, with a third still a distant possibility. After cancellation, the show was syndicated on the PAX cable channel; in 2002 it settled into regular syndication on the Hallmark cable channel.

In keeping with TV's 50–minute episodic conventions, *Dr. Quinn's* storylines were generally not complex. Most weeks, the heroine did battle with a new form of evil, whether this was sexism, racism, gambling, alcoholism, spouse abuse, environmental pollution and so on. Evil was usually personified in the form of a particular individual who was routed or reformed by the end of each episode. The show was often touted as an example of old-fashioned, conservative "family values," yet a closer look reveals a fundamentally liberal point of view, with its feminist message, sympathetic (if stereotyped) Native Americans, and rather ambivalent attitude to organized religion. The show taps into a plethora of American cultural icons that have become part of the folk and popular cultural fabric, from the pristine frontier to the noble savage, from the power of progress to the American melting pot.

The DQMW-L discussion list has existed since 1994; since the cancellation the membership has declined from a peak of over 800, but in 2002 it still has several hundred members. Most are women, but the list includes some men. In addition, there are numerous spin-off Web sites and groups, devoted to individual cast members, fan fiction, and so on. The listers clearly disagreed with the critics' view of the show, and took it to their hearts. Furthermore, they took the list to their hearts; for some people, the list became more important than the show itself, and this is where interesting questions of community arise. Throughout the chapter, I will make liberal use of the list members' voices to illustrate my points, but it is important to note that even with fairly extensive quotation, the comments chosen represent only a tiny fraction of the electronic discourse on the list.[1] Unlike message boards, such as the soap opera groups studied by Baym (1997), e-mail lists allow members to write extremely long messages; I have seen contributions of up to 2,000 words on occasion.

THE VIRTUAL COMMUNITY

Disquiet about the role of media in destroying personal connections is nothing new. Nineteenth-century critics warned about how young women were spending too much time in the unhealthy, solitary company of cheap novels (McFarland 1985). Later, writers such as Meyrowitz argue that "electronic media destroy the specialness of place and time" (1985, 125). Likewise, many critics vehemently deny that "community" can be achieved on the Internet. Doheny-Farina (1996) argues that "a community is bound by place, which always includes complex social and environmental necessities. It is not something you can easily join.... It must be lived" (p. 36). He goes on to say that we must resist the development of virtual relationships or "we risk the further disappearance of local communities within globalized virtual collectives of alienated and entertained individuals" (p. 36). Doheny-Farina is not alone in his disquiet. Stoll (1995) argues that "by logging onto networks, we lose the ability to enter into spontaneous interactions with real people" (p. 43). Anonymous writer "humdog," (1996) writes angrily about the illusion of the Internet as a communal place: "So-called electronic communities encourage participation in fragmented, mostly silent, micro-groups who are primarily engaged in dialogues of self-congratulation. In other words, most people lurk; and the ones who post, are pleased with themselves" (p. 40).

DiGiovanna (1996) is contemptuous about people who spend time on the Internet, especially the most "pathetic" of all, those who create personal home pages in a desperate attempt to "feel like they count" (p. 447), a characterization that echoes notions of the fan as someone who needs a life: "This is what the Net does, in the form of cross-quotes and recycled postings and anonymous remailers and forgeries—it makes itself into an entity which has its own validity while it erases the identity of those who claim to be part of it..." (pp. 456–457).

As I see it, these authors are only seeing half the picture. This is not to say that the Internet is the panacea for a lost sense of community, or that it can replace "real" communities. In fact, I am suspicious of arguments that claim the Internet *is* intrinsically either constructive or destructive of community. Rather, it is a new tool of communication that can be used in many different ways, just as other forms of communication can. It is certainly changing social interaction for many participants. If culture is indeed constituted through communication (Carey 1989), electronic communication may well be changing the nature of our culture. But how that change is manifest is not always predictable; single individuals use the communication possibilities in many different ways. Debate about the value, or even the existence of virtual communities has tended to become polarized into an either/or argument, and I am not attempting to offer a blanket endorsement of electronic communication. People *do* spend more time than ever interacting with machines rather than people (Putnam 1996), and this reality has been blamed for social ills from alienated teenagers to disintegrating families. The prevailing anxiety about the breakdown of deeply-rooted personal ties is articulated by San Francisco *Examiner* columnist Rob Morse, who writes "I have no friends. I have colleagues, neighbors . . . passing acquaintances . . . I have a diaphanous social network of people I . . . often have stimulating conversations with, but whose names I keep forgetting" (p. 15A). There is a very real need for a sense of belonging, which is widely expressed, even as we embrace with fervor the technology we seem to believe is destroying our humanity. It seems important, then, to look more closely at phenomena such as cybercommunities, to see if they do have a potential that critics often dismiss.

Most of the critical vitriol about cyber communication seems to be reserved for anonymous chat rooms, "drop-in" kinds of boards, places where the communication is casual, sporadic, and very often masked. It is hardly surprising that these offer little sense of community; one would not expect to find community in a dark bar where people drop in and out wearing masks, either. But there are other kinds of activities on the Internet, including the one I am looking at, the e-mail discussion group. Unlike electronic chat rooms and bulletin boards, e-mail discussion lists allow for (although do not *necessarily* produce) a membership who exchange long, detailed postings that take the form of episode reviews, extended discussions of issues raised in the program, as well as shared gossip and speculation about favorite actors. Some members of the group spend substantial amounts of time each day participating in the discussion; it is not casual. And although much of the academic debate on cyberculture tends to assume that most virtual groups function anonymously, the DQMW list members overwhelmingly use their real names and addresses.

Maintaining a community, as the critics themselves suggest, takes work. Rheingold, an enthusiastic proponent of the possibilities of virtual communities, writes that "Communities can emerge from and exist within computer-linked groups, but that technical linkage of electronic personae is not sufficient to create a community" (1996, 420). In other words, just being subscribed to a list does not result in a community, any more than just living in a neighborhood makes that a community. Members of DQMW-L see themselves as a community, and they work at it diligently in several ways, through a range of strategies and activities that help them define who they are, sometimes by making distinctions between themselves and other groups. As I shall argue below, electronic communities are not identical or interchangeable with more traditionally-understood, place-based communities, but they may indeed provide a "sense of place" to their members, and fulfill some of the functions of other types of communities.

Nurturing Community

Members spend time and effort welcoming and nurturing new members, and commenting reflexively about the community. It has become a tradition in January of each year to post self-introductions, giving newcomers and established members an opportunity to talk about themselves and reflect on their experience with both the show and the list. For example, in early 2002, a long-time member writes:

> I've been a DQ fan since I was 14 years old. It doesn't seem like it's been nearly eight years . . . I also write fanfiction for the series, which I began doing at 15. Thanks to the encouragement of people such as M. and B., I kept on writing and it has developed into a career. What drew me to DQ was the historical elements that captured my interest and the strong, dynamic women . . . When my parents divorced, the show was my anchor. It kept my belief in finding love alive and kept me hopeful and optimistic. I doubt I would had gotten through that time without the show. Not just the show though . . . That year, I joined DQMW-L . . . This list has some of the greatest people I have ever met in my life, and some of the powerful influences as well. Even through darker periods, it was the people that kept this list together. So many people supported me though my teenage years, I wish I could do something to thank all of them . . . The debates on the list taught me how to observe other people's side of the argument with patience and understanding, something useful in journalism. This list is truly a great place.

Older members often jump in to welcome newcomers and draw them into discussion. Here, for instance, are some moments from a thread that developed in August 1997. A new member has just posted a long message that served as what has been called her "coming out" narrative (Clerc 1996) as a DQMW fan, and explains what the show means to her, a common initiatory feature of fan-based networks. She is nervous about exposing so much of herself to this anonymous group. Many members reassure her:

> I also want to say how privileged I feel that you felt you could share something so deeply personal about yourself with us . . . we are truly a "family"—a large, caring community which spans the globe . . . We exult in each other's joys, and cry over each other's

sorrows. I speak here from personal experience, having just lost my mother two months ago. In the days following her passing, the outpouring of love and support I received from this List was one of the things that sustained me most; and, as time has passed, has been what has aided me most in my "healing" . . . In the weeks and months to come, I hope the love and warmth of this List will sustain and shelter you as much as it has me. Welcome to our "Circle."

Others take on the discussion, developing more metaphors to describe the list, such as "porch" or "quilt," as they address the new member:

I hope that you continue to gift us with your posts . . . You are also about to discover one of the warmest, most giving groups of people you could have possibly stumbled across . . . Cyn's Front Porch is a place where people will hold you when you need holding, cry with you, give you strength and courage, offer words and prayers, and simply share the pain and the triumphs of their own lives. We go through deaths and births and graduations and weddings—opening nights, publications, cancer scares, and every other human condition—together.

Another writes:

I was thinking again this morning of the metaphor we have used in the past of the list being like a quilting circle where each of us comes to share and we bring our own pieces of fabric that make us unique; sewing the different offerings together to make something special and memorable . . . I suppose it is why I love the list; because it is full of . . . people whose experiences are not all the same; who learn to . . . work to treat each other like friends . . . rather than cyberstrangers who greet each other with suspicion or isolation. It is why, for me, even when I am disappointed with the show itself the list remains important to me . . . I think it so fitting that a show about a Pioneer Woman would promote a kind of Cyber Quilting Circle . . . It is hard to "know" people in this faceless medium; it takes special ears to hear voices and snatches of each other's lives but in talking about Dr. Quinn I think that is what we often do . . . In a way cyberspace is as bleak and impersonal as a bare soddy with primitive furniture out alone on the prairie. It was a functional dwelling without personality . . . I guess I think what we have here is the same sort of process at work; we fit and piece and bring warmth to the starkness.

Although not all messages are so eloquent or self-reflexive, the rhythm of the list regularly includes such moments when the members attempt to define what their "community" is all about. The literate, rather than oral, nature of the discussion, relying heavily on the written word (unlike the more spontaneous "real-time" character of a chat-room) actively encourages such introspection. The "quilt" metaphor seems to have been an especially powerful one to the community; in fact it has taken on concrete reality. In 2002, coinciding with the 10th anniversary of the conception of DQ, a member of the list initiated a project to have members create a tribute quilt, with each sewing a square that captured her feelings about the program. The finished quilt included 56 squares, ranging from fairly simple renderings of words and patterns to quite complex and artistic compositions. It was presented to DQ creator Beth Sullivan, but it lives on as a special display on the official DQ website. Its existence consciously evokes the ideal of the list community as a collaborative, female group, articulated several years before:

> Quilting was a way to turn bare rough houses into homes, satisfying women's hunger to create and have beauty . . . My best advice for new people . . . is to "bring lots of material." Bring more pieces; talk about your blocks, be they fanfic, general observations, history, humor, pieces of yourself, keep offering to the quilt; allow your voice to become known; and never assume ANYONE thinks your fabric is not welcome . . . Quilts need lots of pieces and quilters can't hang back; they have to get in there with their needles and thread and conversation . . .

Establishing Rules

A frequent criticism of electronic communication is that it is so open, unstructured, and anonymous that it encourages aggression, and lack of concern and respect among those communicating. Once again, this is something that DQMW-L members actively work to avoid. Most e-mail lists have rules about civility, but breakdowns in those rules are among the most pervasive problems (Coates 1998; Dery 1994). DQMW-L has a conduct code, sent to members when they subscribe. The list is remarkably free of flaming and insulting posts, mostly because it has been long watched over by the list-owner, Cynthia, and she and her successors are overwhelmingly supported by the members if any step out of line. "Disciplinary" actions are occasionally taken, such as removal from the list, a last resort that has been employed only when disruptive members have used personal invective, profanity, or racist language. It has been threatened when members have contravened established rules, such as the prohibition on sending jokes, chain letters, and other non-relevant messages. However, these decisions are open for discussion; a new young member once sent a message that contained a jokey graphic design that took up inordinate amounts of band-width, and was promptly expelled by Cynthia. However, other members spoke up for her, citing her neophyte status, and her remorse. She was re-instated, made an apology to the group, and went on to become a regular poster. This example points to the negotiated and flexible way rules are maintained, as established members use their status gently to remind members about the rules against irrelevant posts, quoting previous posts, and attachments (which can harbor viruses and cause problems for some members).

In addition to the maintenance of more formal rules, there are frequent discussions about general conduct on the list, and suggestions about how to keep it a civil place. Thus, for example, an established member writes after a brief flurry of rather hostile posts:

> I think it would behoove all Newcomers to listen for a bit, before plunging in with both feet. I listened and asked and learned from my List Elders for quite a while before I felt I had earned the place to speak from a position of having paid my dues. I would hope others would think about doing the same. There is more to learn here than anyone would imagine, when you first set foot on this front porch. . . . I guess I'm asking for a little caution, a little patience, a little prudence.

Another adds:

> (It's) not because Ye Olde Geezers have anything more important to say than new folk but because one of the things that makes this place special is the *learning* of voices; the circle that allows us to "recognize" each other. . . . Many lists never shoot for that; it takes work and grace and belief in the list community itself.

As Rheingold writes,

> The ultimate social potential of the network. . . . lies not solely in its utility as an information market, but in the individual and group relationships that can happen over time. When such group accumulates a sufficient number of friendships and rivalries, and witnesses the births, marriages, and deaths that bond any other kind of community, it takes on a definite and profound sense of place in people's minds (p. 420).

This sense of place, and feeling of comfort and recognition, seems to be the goal of DQMW-L members. Like the soap opera fan groups studied by Baym (1997), the DQ list members fall into recognized categories, such as "geezers," "newbies," and "lurkers." "Geezers" carry some authority in the group, and "norms implied by and embedded in their messages carry a good deal of persuasiveness" (Baym 105). The existence of this hierarchy also adds a sense of structure and continuity to the group. Geezers provide a kind of "institutional memory," sometimes reminding members of previous discussion threads in related topics, or occasionally posting "reruns" of "classic" messages, which are often appreciated and discussed anew.

Discussion of Issues

A great deal of the discussion on the list involves detailed review and discussion of episodes as they are shown; in fact this is ostensibly the main purpose of the list. The rhythm of the discussion follows now established patterns: The first round of messages will be "reviews": what was good about the episode, what was wrong, and so on. This has continued, even though the show is now only in syndicated re-runs, and many listers have seen each episode numerous times. After that, there begins a series of discussions that generally reflect on larger themes, and involve consideration of personal issues. For instance, there was extensive discussion of an episode in which Sully aided Cheyennes in their escape from the nearby reservation, thus placing himself outside the law. This led to detailed debate about rights and obligations: Did he owe his loyalty to the Indians, or to his family, who ended up alone when he became a fugitive? What about his obligation to follow the laws of the nation? Do higher moral laws negate such an obligation? Lawyers weighed in with legal opinions, historians contributed ideas on how the events tied in with Native American history, and so on. We can follow the development of a particular discussion to illustrate this process.

On April 6, 1997, an episode titled "The Body Electric" was shown. In this episode, the poet Walt Whitman comes to Colorado Springs, where he is welcomed until it becomes generally known that he is homosexual, and his young male lover arrives in town. Then he is shunned by most of the townspeople, and Dr. Quinn has to grapple with her own dislike of his sexuality. Eventually, she stages a poetry reading, which is boycotted by almost everyone in town, with the exception of her family. As I have mentioned, while the show is known for cozy "family values," its overwhelming perspective is liberal, a perspective shared by most on the list (at least while the show was in its original run), so initial response was quite favorable. There was extensive discussion of Whitman's life and work, of the motivations of the various characters, of the derivation of the term "Nancy-Boy," used by a character to describe Whitman, of historical and Native American attitudes to homosexuality, and so on. Some typical comments:

> It makes me feel good that we've had two episodes that have ended with Whitman's poetry being read. . . . I always feel enlarged, re-sensitized to the world around me when I've dipped into his work. That he was homosexual was never an issue . . . That's what came through in tonight's episode, and I was proud of Mike for being able to overcome her initial shock and see through the fear to the treasure of his words and humanity.

Others, however, began to disagree, both on grounds of historical accuracy and personal taste:

> Being the time period it was . . . I found it hard to believe that the church/Christianity didn't come in to play at all. Dr. Mike consulted her medical journals, but never her Bible . . . I know others on this list may feel differently about this issue . . . but I think that, based on Michaela's previous standings, that she would have at least consulted it.

The discussion took a different turn with a contribution from one of the few active men on the list at that time:

> This latest episode . . . was of course totally unrealistic and pre-dictable. Homosexuality as a way of life has NEVER been accepted by the general public until the last few years when the homosexual community have forced their agenda on society . . . [Producer] Beth Sullivan has shown how she . . . is using Dr. Quinn to promote her homosexual-loving agenda . . . If Beth really wanted to show the real situation, she would have the reverend speaking the truth—that is, homosexuality is immoral and totally against Christian teachings . . . Promoting homosexuality in a family show is a terrible thing to do.

This could be construed as "flame-bait," but the list treated it quite calmly. A regular poster comments:

> J., while I disagree with the homosexual lifestyle on a religious basis, I feel the need to point out to you that the Bible also says that we should love ALL PEOPLE, no matter what they do. You can love the person, even if you hate the act. And yes, I DID disagree with Sully's view in this episode, and I am not afraid to say that . . . I am not saying that you have to like homosexuality.

All I'm saying is that if you really look deeply into the words of Christ, you'll find that he wants us to love everyone . . . Hate never solves anything. JMO.

J. counters with more, saying, "IF you want to start quoting verses, there are several verses in the Bible where it states that 'Homosexuality is wrong' in plain English. Remember Sodom & Gomorrah." Other listers respond:

Dear J., I think that you are probably young, male and white which gives you a lot of privileges which you may not appreciate yet and which others don't get to share. I do not agree with your religious point of view but I would fight for the right for you to state it. I hope that maturity will temper your anger and that you will realize that love is what matters above all else.

This and other similar messages brought out another male lister in sarcastic mode:

J. : S. is absolutely right. You ought to be ASHAMED of yourself for being young, white, and male. Heaven knows, there are too many of you over-privileged snots running around, spouting your ill-thought and irresponsible opinions to anyone who will listen. The nerve! . . . S. Bravo! Welcome to the club! You've learned the Secret Handshake: when in need to establish rhetorical authority, be sure to reference the antagonists' age, race, and gender—they are the only things that matter in this world, right? Say, I know where I can dig up some barbed wire real cheap—let's go build a concentration camp for young, white males in your backyard. I'm sure we can get some volunteers from the List to serve as guards.

A wave of objections follows:

You have every right to your opinions, you even have the right to express them on this list. What you don't have the right to do is to hurt even one single human being. I am terribly afraid that whether you meant to or not you already have done just that.

Another then asks to have J. removed from the list because "Most of the list values the community and is hurt by actions that attack the community." A "geezer" on the list tries to calm things:

With a community as diverse as ours it would be unreasonable to think we would all agree on everything. However, it is essential that a degree of civility, respect, and tolerance for divergent viewpoints, cultures, backgrounds, and lifestyles be maintained. Deliberately inflammatory or denigrating remarks should be avoided.

Later, she continues, quoting some sections of the list rules:

> Internet discussion groups are an experiment in electronic community development. Remember that the written word is understood differently than the spoken word so please edit your posts for clarity and remove inappropriate emotion.

List-owner Cynthia finally weighs in with a similar message, no one is removed, and eventually the debate dies down, although not before one lesbian lister offers a long personal response ending:

> The overwhelming support I have witnessed from this list for the sensitive handling of [the] situation has brought tears to my eyes . . . and thank you to the listmembers who HAVE NOT shown their support, it offers other individuals an opportunity to see what some gays and lesbians endure . . . My sexual orientation is lesbian. My lifestyle is that of an almost 40-year-old female who looks in the mirror daily counting the new gray hairs. Who rises out of bed in the afternoon to work a night shift . . . How much different is my lifestyle than the lifestyles of others on this list?

From this greatly condensed summary of several days of discussion, we see how list members use the show as a catalyst for discussions that go far beyond the specificities of the program itself, allowing people across a wide geographical and social spectrum to communicate. It may not be the same kind of communication as that experienced in daily personal interaction, but it is difficult to dismiss it as meaningless. As a list member wrote in April 1999, following another intense discussion:

> [We have] discussions like these, and people have the absence of knowledge to suggest we're just a bunch of cyber-loonies! I haven't had a discussion this involving at work or at home in weeks . . . Think of it, someday someone will analyze this list the way we do the show.

As a regular lister writes, "Discussions that mirror the fictional debates of folks in our favorite TV show can be stimulating and fascinating. Sometimes, public discussions are a metaphor for our experiences of living and faith."

A Female Space

One feature of the Walt Whitman debate is probably apparent. The most aggressive and perhaps "inflammatory" comments are made by men, while the women work to smooth things over and be inclusive. This turns out to be very typical of Internet communication (Balsamo 1994; Kramarae 1995; Warnick 1999). Once again, it is most likely not the Internet in itself that creates this, but rather a well-established difference in gender communication styles that results in a situation in which "electronic discussion lists are governed by gendered codes of discursive interchange" (Balsamo 1994, 141). As many scholars have shown, female communication styles tend to be collaborative and inclusive, building on what others have said, making the DQMW list's "quilt" metaphor seem quite appropriate (Tannen 1993; Maltz & Borker 1999). Male styles tend to be more confrontational and interested in making decisive points that "win" an argument. The exchanges on Whitman illustrate this well. Many scholars of the Internet are arguing that the argumentative male style is the dominant form on the Internet; Coates (1998), in a discussion of gender issues on the Rocklist discussion group, argues that women were effectively cut out of the group: "Rocklist . . . not only replicates the gender dynamics of the rock formation in general, but, as an Internet forum, also replicates the social dynamics currently operating in nascent cyberspace" (p. 81). DQMW listers see this too, writing often about how important it is to maintain a space that, although not necessarily only female, supports "female" modes of communication. Indeed, Clerc argues that "Media fandom wouldn't exist without women because more women than men do the communication work necessary to forge and sustain community" (1996, 78). This awareness of the "female difference" is expressed clearly in comments from a list member who compared the DQ list with another e-mail list for fans of the work of J. R. R. Tolkein, which was mostly male, unmoderated, yet very civil:

> Discussions ranged from in-depth examinations of medieval philosophy in the works, to what or who exactly was Tom Bombadil . . . In many ways, this sounds much like our group. But . . . the openness we have was not encouraged . . . the personal interaction, particularly in times of joy or sorrow, that we enjoy here was not even conceived of there . . . It was a designated space for a designated purpose . . . much of the discussion, while not combative, was—demonstrative, maybe. In the sense that people were demonstrating why they thought what they thought. Used material from the books to prove their points . . . Ties to self and the larger world were not really part of the discussion. . . .

A 1998 exchange on DQMW-L underlines this point. In April of that year, an episode aired in which a regular character, Hank the barkeeper, set fire to a Chinese immigrant settlement, fearing an epidemic. Hank is a rather inconsistently-written character, who swings between extreme bigotry and rather tender moments, and he has a substantial fan following. During the discussion, people questioned his act, saying it showed a regression in the character. One of his major fans writes:

> I'm making the assumption that Hank was trying to save the lives of the town—since he believed there was another epidemic starting . . . It all depends on what you believe the motive was for Hank.

A male lister disagrees, and argues his points on "the facts":

> You seem to be asserting that whether or not Hank committed a criminal act is a matter of opinion—yours versus mine. It's not. Hank *did* douse the Chinese camp with kerosene . . . I'm frankly baffled that you defend him.

Another member immediately begins "communication work," saying she has enjoyed the disagreements:

> It's what makes this group so special to many of us. G. tends to focus on the wounded psychological aspects of (Hank) while M. has brought up more logical/legal issues. Both are valid, but logic seldom changes emotion and emotion seldom influences logic . . . Some people are judgers, others perceivers, some thinkers, some feelers. Neither is right nor wrong, but they express how differently people approach problems.

Although men are not excluded on the DQ list, they are in the minority, and soon learn to adapt their communication style to fit the predominant mode, or else they drift away, as many have done. DQMW-L *is* a women's community, as many listers explicitly say, while often deflecting possible charges of "Pcism" with their light-hearted mentions of the "F-word"—Feminist.

The female identity of the list was made far more explicit in the months following the cancellation of the show in early 1998. The cancellation was unexpected; although DQMW's ratings had been sliding, it was still winning its Saturday night time slot, and attracted a regular audience of about 12 million. In media explanations for the decision, CBS president Les Moonves explained that the network needed to attract young, urban male viewers, and that DQMW viewers were "too old, too rural, and too female," words that became infamous among the DQMW-L community.

Industry wisdom is that while middle-aged women have spending power, they are not so easily swayed by commercials, and are thus an undesirable demographic for advertisers. CBS followed this through with a determined effort to attract young men, primarily with NFL football. To add insult to injury, the network replaced *Dr. Quinn* with *Martial Law*, a cop/martial arts series. Although it attracted fewer viewers than *Dr. Quinn*, it apparently reached the desired demographic, and was briefly deemed a success. It faded fast, and was cancelled in early 2000.

After the cancellation, DQMW-L listers launched an immediate attack against CBS, demonstrating how effective the Internet can be in mobilizing people quickly and efficiently. A "Save Dr. Quinn" Website and committee emerged, spawning letter-writing campaigns, rallies, and phone barrages. A measure of the trust among the group was that $13,000 was raised to place ads in such publications as *TV Guide* and the *Hollywood Reporter,* all the money being sent in by listers to organizers they only knew through the list. CBS acknowledged that this was the biggest and most persistent fan campaign they had ever seen, and eventually agreed to make at least one *Dr. Quinn* TV movie. It was shown during the May sweeps period of 1999, and was followed by a second, in May 2001. Currently, fans are maintaining an active Internet petition for a third movie.

During the campaign, the listers became much more assertive about their identity as women, who felt marginalized by the power of CBS. The "F-word," was used more frequently. As a list member wrote in early 1999:

> Most of the criticism of DQ as hokey and sentimental sets my teeth on edge the same way Les's comments about the audience being "too old, too rural, and too female" did . . . I think DQMW presents a unique threat because it has invaded the space of the western, and brought a woman's voice into the bastion of male fantasy and female silence.

Indeed, a sense of being under threat or marginalized in some way can be a powerful force in building a sense of community, and in its post-cancellation phase, DQMW-L developed an even more integrated voice as a "female space."

Building Community Through Fantasy

The bulk of activity on the List consists of the regular postings related to the show, its actors, and off-screen activities. In addition, List members participate in the many Internet sites about individual actors, and write and read fan fiction, much of which is initially posted on the List, but eventually added to more permanent Internet sites. Some Internet sites have been more fanciful than the usual fan pages, such as the now defunct "Goodbody" site, started by a group of Joe Lando fans who had become known as the "Goodbody Sisters." Jasmine Goodbody, for instance, posted for several years the "Love and Lust" reviews of episodes in which she lampoons the show's conventions affectionately, playing up her supposed uncontrollable lust for Lando, as she clutches him to her "ample bosom." Jenson nicely describes the stereotype of the fan: "If she is female, the image includes sobbing and screaming and fainting, and assumes that an uncontrollable erotic energy is sparked by the chance to see or touch a male idol" (p. 15). Jasmine Goodbody, and others on the list, consciously mock this stereotype, declaring themselves to be "swooning and sighing," and overcome with desire (while at the same time acknowledging that a major part of the attraction of DQ is indeed the male leads). The same stereotype of the lustful female fan is both celebrated and mocked in the list sub-group known as "Hank's Hussies," whose object of desire is Hank the bartender, played by William Shockley. The Hussy Website, for instance, features computer merged photos that place various "hussies" into fantasy situations with Hank.

Common understandings of "fantasy" tend to perceive it as a solitary process, in which individuals develop their unspoken desires alone or in strictly intimate situations. Yet, as the DQ list shows, fantasy can also be a collaborative project. An activity that comes and goes at different times on the list is a fantasy trip known as "Going through the stones." It evolved in the early days of the list, and was inspired by Diana Gabaldon's science fantasy *Outlander* books, in which people travel into history through magical standing stones. List members began creating virtual adventures, posted as e-mail messages, in which they travel back "through the stones" to 1870s Colorado Springs, meeting both DQ characters and other list members on the way, melding past and present, "real life" and the TV narrative. Thus a 1995 Mardi Gras trip, written by the then "High Hussy" begins:

> I always think of Hank as being from the Big Easy . . . now if he can't celebrate down South in 1995, maybe it's time for a trip through the stones . . . <grabbing green and purple gown, gold boa, and feathered mask . . . > (Back in 1870) . . . Goodness, these trips are getting easier and easier. I guess it's like what they say about childbirth . . . the first one takes forever, the rest slide right through. HANK was glad to see me. The saloon business has

been slow...he had some outrageous memories of Fat Tuesday and was more than willing to try it in Colorado Springs. In fact, after a trip to Loren's, I was surprised how READY this town was to PARTY!!!...At sunset I made it back to the saloon. Hank had it decorated and had brought out the GOOD stuff. I had left my gown in Myra's old room; and when I opened the door, I discovered roses all over the bed and the room reeked of French perfume...Within the hour there was a knock on the door. I opened it to discover the Bacchus King...a white sequin suit, cape and feathered hat with a gold mask. He held out his hand and bowed, but the sexy sneer gave him away...IT WAS HANK!!!

In a reply to the whole posting, another member invokes one of the list's characteristic tropes—the melding of modern technology with nostalgia for a romantic past, which is then picked up in High Hussy's answer: "Hussy, how do you get through the stones? I can't seem to figure this out. Do you need special software??? ;-)" High Hussy replies:

Well, there is no special software. In fact, I have found openings to inner space in almost every town. The first trip was during an AOL Chat. Actually, I had just gone to bed when I felt Leigh grabbing my hand and pulling me along. The next thing I knew was I was standing on a dirt street with Hank's Saloon in front of me. Well, I didn't waste any time...running inside. BUT MY AOL FRIENDS LEFT ME!!! I spent a week there...The best times for stone crossing are at dawn or summer solstice. You can cross at other times...but you can experience pain. <Well known stone/time warp sites>:
*Culloden in Scotland: made famous by Claire Randall-Fraser (*Outlander* series).
*The jungle room at Graceland: Elvis went back to the 50s to see his mom.
*Stonehenge: best for quick trips to see King Arthur and Lancelot.
*Easter Island: the faces must be facing each other otherwise you don't return.
*The Senate subway: statesmen enter to be replaced by politicians.
*Algonquin...in the tea room at 5:30 P.M. on Friday: Dorothy Parker will usually show up. K., I checked my map...the nearest one to you is room 416 at Grand Hotel on Mackinac Island. Safe crossing.

In 1997, an elaborate "stones" trip developed over several months, with numerous members contributing. It followed outrage on the list about the adding to the cast of former *Dukes of Hazzard* star John Schneider, who was widely perceived as usurping the place of Joe Lando's Sully. In retrospect, it seems clear that the decision to hire Schneider was the first sign of CBS's desire to broaden the audience, as media publicity claimed that his "man's man" persona would bring in young male viewers. The core audience was not impressed; Schneider was seen as a lightweight actor, associated with a "trashy" former series, and there was much vitriolic comment on the list. To lighten things up, a member suggested a military operation in the form of a "stones" trip back to Colorado to remove Schneider's character, Daniel, and send him to the "Black Hole," where written-off characters disappear. Volunteers for SAFROD (Secret Agency for Removal of Daniel) would need appropriate clothing—"Calico dresses would work . . . I don't suggest any floozy dresses because we don't want to get pulled aside to go to work at Hank's establishment"—and appropriate demeanor–"we must practice our shuffling and pacing . . . You have all seen how the people nobody knows in CS shuffle up and down the sidewalks of town. I think when you get to one end of town you turn around and go back the other direction. Practice this move—it will promote your personal safety." The operation developed into a long series of fantasy narratives spanning four months and thousands of words, in which the troops, taking on special identities of their own ("Modern Major-General," "Special Agent," "Captain Anthro"), fanned out across Colorado Springs and Quinnland, posting accounts of what had happened as the "soldiers" infiltrated the town looking for Daniel. The stories blended their "real-life" identities with DQ characters, affectionately playing on the show's conventions and characters, their knowledge of each other, and their personal passions, amid an eclectic patchwork of cultural references, from *The Avengers* to Native American history to ginsu knives. Many members posted individually, while others wrote elaborate collaborative pieces, and the original leader regularly brought the narrative threads together and helped move them along. Many of the narratives were wildly inventive and funny; unfortunately, it would take a separate chapter to do justice to the creativity that SAFROD engendered.

Fuller and Jenkins comment on the many ways the Internet is conceived of as unexplored land: "Virtual reality opens new spaces for exploration, colonization, and exploitation, returning to a mythic time when there were worlds without limits and resources beyond imagining." (1995, 58). Clearly the stones trips reflect this same conception of the Net, although a collective one. They are an activity that was made possible only by Internet communication, and by the particularly collaborative, female style of groups like DQMW-L.

A Sacred Place

In reflecting on their community, DQMW listers are aware that however virtual their group may be, for most people community means place. Unlike anonymous groups, who often revel in the sense of placelessness that the Internet can provide, listers constantly invoke metaphors of place. Even though the list founder, Cynthia, is now largely absent, her name lives on in the "front porch" symbol used repeatedly. Newcomers are invited to "pull up a chair on Cyn's porch," re-emerging lurkers "stroll back round to the front porch," and so on. The chosen metaphors, such as porches, lemonade, quilts, and rocking chairs, consciously evoke a bygone era, and that era is place-bound—Colorado Springs in the 1870s. The fantasy stones trips involve travel to that place, and just as significantly, the list has another geographical focus. This is not the actual Colorado Springs, but rather the fictional set. *Dr. Quinn* was filmed in a public place—the Paramount Ranch in California, located in a National Park, and managed by the park service. The set was essentially open to the public, and many listers traveled there on a sort of pilgrimage. Members would discuss travel plans on the list, and post "Pranch reports" about their trip, mentioning which actors and crew they met, and what it felt like to be at the sacred place. The March 1999 resumption of filming, for the upcoming TV movie, was greeted with fervor, as list members planned what might be their last opportunity to visit. Messages were exchanged with tips on hotels, good vantage points, and correct behavior on the set. Photos are posted on websites, and listers arrange to meet in person at the Pranch. Listers are careful to warn others about good behavior, and the need to ask responsibly and not like "crazed fans." The "Pranch visit" is clearly the ultimate goal of many listers, as one suggests:

> I would have never known that I could visit the set if it were not for the list. I can't even describe what it felt like the first time I made my first Mecca to the ranch. Although I made a few more, after that they were just as exciting as when I made my first trip across the bridge to Colorado Springs.

Hardy and Kukla (1999) point out that television theorists have usually assumed that a program's setting merely "provides a . . . backdrop for the events it portrays" (p. 185). They disagree, arguing that for fans of many shows, a sense of place is crucial. *Star Trek* fans have a great interest in the physical layout of the Starship Enterprise, and "a primary text-making event" is "an active construction of the ship—which the audience experiences only through fragmented and separate visual images over time—as a fully imagined and unified aesthetic space" (p. 180). Likewise, DQ fans use both Colorado Springs and the "Pranch" as imaginative centers for their activities, while being in no danger of losing touch with reality, as critics often assume. "Our imaginative entry into narratives . . . does not falsely collapse the distinction between self and text. Instead, it highlights the active, dynamic process by which narrative possibilities in relation to self and community are created, transmitted, and rejected" (Hardy and Kukla 1999, 184–5). While not all TV shows have quite such a developed place identity, many do, as suggested by Couldry (2000), in his discussion of visits to the set of Britain's *Coronation Street,* and in the rise of tourist activity centered around media productions.

Aden (1999) writes about how simply participating intensely in popular media stories "is a ritualistic journey of the mind to spiritually powerful places where a vantage point . . . affords us a reassuring view of an imagined promised land" (p. 8). Indeed, the immersion in DQ discussion, plus the fan-fiction and fantasy stones trips, do facilitate that "journey of the mind," a journey that becomes even more real when supplemented by the real-world journey to the symbolically-important Pranch. As Hills (2002) writes, "cult geographies allow fans to fantasise extratextually 'inhabiting the world' of the text in precise and detailed ways" (p. 157), and both the virtual and real journeys undertaken by DQ fans do just that. The sense of place has, if anything, grown more intense since the series itself folded. In particular, the DQ official website, whose management was taken over by a DQMW-L member, has become a very elaborate structure. It represents the fictional Colorado Springs, and features different "buildings," such as the saloon, church, mercantile, and so on; each features information both about historical context (the clinic, for instance, has documents on nineteenth-century medical practices) and contemporary fan resources (videos and book available at the mercantile).

THE INTERNET FAN

My ethnography of the DQMW-L list suggests to me that Internet communities do indeed exist. Like other groups discussed by Rheingold, DQMW listers explicitly describe themselves as a "community" and work hard to explain and defend their use of that term. As one lister writes, "I count this List as one of the great finds of my life! It has given me lifelong friendships, which isn't easy to accomplish once one becomes established in home and career."

This is not to suggest that the Internet builds community by its very nature. As Gurak (1997) writes, "it is important to move away from generalizations about life in cyberspace" (p. ix), and move toward looking at ways in which people in specific circumstances use the technology. Building community is not necessarily the goal of electronic communication, and we should not necessarily wring our hands when it does not happen, nor should we assume that a person's online activities define his or her existence. For example, part of the excitement of anonymous forums is the role-playing and multiple identities they allow (Curtis 1997; Turkle 1997). Many computer users see these as liberating, allowing unheard-of opportunities for creativity, fantasy, and fun. Does this mean that they are alienated, rootless individuals, who have lost the need for human ties? Maybe so, in some extreme cases. But other individuals participate in technology and in "real life" in multiple ways. Members of DQMW-L write that they use the technology for different purposes at different times. Some participate in anonymous chat rooms, some are members of other e-mail groups that they join for informational or professional purposes, but from which they do not expect "community." Others stress the fact that their daily electronic communication is just one element in their lives.

Internet communities, like place-based communities, do not just happen. They develop in response to particular circumstances and to the needs of a particular set of individuals. A striking thing about the DQ list is the level of trust and openness that has grown up in it. Critics bemoan the implications of electronic anonymity and role-playing for traditional notions of communication. Yet DQMW-L points to the fact that Internet communication does not have to be anonymous. Some listers like to use a "pen-name," but none hide their real identity. There is an online, voluntary directory, where members post their names, states, and e-mail addresses, and give details about themselves. Listers meet in regional get-togethers, and visit each other. They warn each other about the dangers involved in anonymous chat rooms, and compare their community to the otherwise bleak cyberscape.

My ethnography also suggests to me that we still do not understand all the possible dimensions of being a "fan" and what we mean when we talk about them. Some scholars have critiqued the denigration of fans as obsessive neurotics, and the DQMW listers themselves agonize about that label, as in a discussion about "fandom" that developed in October 1997:

> I've been trying to figure out when a person crosses over from being "a normal viewer who is entertained every week" to a dreaded "fan" . . . Let's see. Synonyms for Fan are: supporter, enthusiast, partisan, booster, addict. When DQ is good, I guess I am a supporter . . . I *hide* the fact that I watch, so I would not call myself a

booster. I also criticize the show a lot, I guess. But can't one be a supporter and still be a critic? . . . I've already gathered that if you dare discuss a show on the Net, you are automatically assigned to the list of weirdo, psychotic, freaky-geek viewers who have no lives . . . It makes me laugh, because the ones CBS seems to be after are like my brother, who sits on the couch all day with a glazed look in his eyes, hardly using a brain cell while he flips through 99 cable channels in his never-ending search for football games.

Another contributes:

What I find most striking is that our involvement in and discussion of fictional characters should be considered at all remarkable . . . While modern media does seem to make the phenomenon of "fandom" seem like something new, I know from my experience that the same type of critical activity is not particularly unique . . . My own prior experience was with the scion societies of the Baker Street Irregulars, devoted to the discussion and study of Sherlock Holmes and the works of Sir Arthur Conan Doyle, encompassing both serious and frivolous participation, including pastiches, the equivalent of fan fiction . . . People often dressed in Victorian costume talked very seriously about Holmes, Watson, and generally had a good time. We all had "real" lives, too.

The DQ listers in many ways epitomize the "fan-as-scholar" discussed by Hills (2002); in addition to their literate, critical analyses of texts, they constantly reflect on both the nature of their fandom and their particular fan community. For instance, independently of my research, "Kathy," an active list member, carried out her own, informal exploration of fandom, inviting members of her two lists (the largely-female DQMW-L, and a mixed-gender list for high-IQ adults) to comment on why they might become "engrossed" in a TV show. She concluded (in posts she shared with other members and later with me) that one common thread among high-IQ members was that they

connect with shows that ask "what if" and depict alternate realities. Science fiction and fantasy rank very high with these people. Others seem drawn to shows that depict the complexity of the human condition—"SOBs who show glimpses of compassion," or "good" characters who are drawn into their darker nature. So far, none of these GT respondents connect to stories that are based in what most would consider a likelihood of possibility. That means historically-based shows like DQ, or contemporary dramas.

Many also "expressed the ability to connect with characters in a way that wasn't possible with real people around them." For instance, one wrote about being a fan of an obscure 1970s British science fiction program, *The Tomorrow People*, which focused on a group of youngsters with special abilities such as telepathy. He writes "In many ways, the show was a metaphor for childhood giftedness . . . Can you imagine living your whole life in a tribe of monkeys, with your very survival depending on them not finding out that you're a human being, a superior creature? That's the sort of training these kids have had."

Although Kathy saw differences between the two groups, she still saw "commonalities that transcend the superficial," most clearly in the importance of fantasy. Female DQ fans "looked at the relationship between Michaela and Sully and saw something that harmonized with the longing in their own hearts. I think the men in the other list are not too far afield, although they express it differently. Many of you expressed that you became drawn deeply into DQ during a particularly stressful time in your lives. These men also found that human connection in the characters—a connection that was missing in the "real world" around them." And members of both lists spoke of the "reinforcement of the cyber-community."

Having established the positive dimension of fandom, Kathy then asks the logical next question: If people find connections in cyberspace fandom, does that mean they lose touch with reality and real relationships? Her interest is sparked by a thoughtful DQ fan, who has developed an abiding interest in a minor recurring character, creating an entire story for him through a website showcasing her fan fictions and critical analysis. In her response to Kathy, she muses on whether her "obsession" is unhealthy. Kathy replies:

> Well-developed relationships in shows and stories screen out the superfluous and intensify or compact personal experience. In that sense, drama is increased and resolution truncated when compared to real experience. If one begins to expect that kind of intensity on a day-to-day basis, then one is doomed to disappointment in any long-term relationship. I imagine there are a lot of people who do become addicted to the drama, but I'm not sure if TV relationships play any part in that. I suspect personal experience . . . plays a greater role. I also suspect maturity is a component in developing a realistic view of the potential of relationships. Just as one puts aside the fairytales of childhood, one bases a mature ideal of relationship through life experience.

Kathy's comparative approach, which attracted great interest on the list, points to intriguing possibilities for more in-depth, ethnographic study of electronic fandom as a new and extremely pleasurable and satisfying aspect of life in a media world, as people engage with different texts and different interpretive communities at different points in their lives, intermingling their "imaginary social worlds" (Caughey 1984) and their relationships with real people.

The "fans" who join an Internet discussion list are a particularly engaged sub-set of the many individuals who like to watch a specific program, and enjoy using electronic communication. Even within that sub-set, the kind of "fandom" varies greatly. Some are deeply interested in the actors and spend most of their time discussing and speculating about their lives. These kinds of fans work hard to get close to the actors, sometimes using organized fan activities as a way to do that. They may follow their favorite performer's career, rather than focusing on one particular role. DQMW-L had members who came to DQ because of Jane Seymour, and continue to hold her as their focus, while others are more interested in cast members in terms of their DQ roles. In either case, devoted fans often develop a sense of a relationship with "their" actors. It may be very internalized and unspoken, as with the woman described by Caughey (1994), who identified in a rather complex way with actor Steven Segal. DQMW-L members were very conscious of the perceived social disapproval of people who might be considered "obsessive" about an actor, and there are frequent reminders about separating characters from their portrayers. Discussion, even harsh criticism, of characters' flaws is acceptable, while attacks on actors are not, as one member learned:

> I learned my lesson a few years back in my first week as a Lister. I had written a fake interview with Jessica [Bowman, who replaced the very popular Erika Flores as Dr. Quinn's adopted daughter, a move that was disliked by many] which made her come off looking like an ignorant airhead. I thought it was funny, as did a select few, but most were fuming and I am grateful for my second chance. The difference there was that Jessica is a human being with feelings and I had no right to make fun of her the way I did. However, Colleen is a character, and when she says or does something atypical, I feel it is my (and all of our) jobs to say something about it ... Sometimes we tend to associate ourselves with the characters, which is only our way of showing our passion for the show, but we really do have to step back and realize that they are characters.

At the same time, members are very open about their personal attractions to particular actors, and will write at length about their favorites.

The chance to meet a DQ actor is the highlight of many fans' lives; chance encounters are described on the list, and pilgrimages to the set are detailed, with particular emphasis on meetings with stars. The various DQ-related Web sites feature pages of photos of fans posing delightedly with the various actors.

And while there is certainly a "star-struck" quality about much of the fan reaction to celebrities, fans are also very conscious of the implied relationship between fan and performer that suggests an obligation owed to fans. The message is clear: "We put you there, and we expect something in return." For the less engaged fan, that something may simply be an expectation of good behavior, as expressed by many of the tabloid readers I spoke with (see chaper 2), who were quick to judge those who put themselves in the public eye and were found wanting. For fans like many on the DQ list, the expectations may be higher, especially after time-consuming and expensive efforts like the "Save Dr. Quinn" movement and the subsequent successes in gaining CBS commitment to the TV movies. For instance, in early 2000, during negotiations about the second movie, Jane Seymour appeared on two daytime talk shows, and made some negative comments about "Internet fans," who had vocally opposed a plan to cut costs by filming in Canada and excluding many of the favorite cast members. Seymour questioned the right of fans to interfere in production decisions (the Canada filming was eventually shelved), and implied that she actually found them a little frightening. A torrent of discussion, much of it uncharacteristically angry, deluged the list, with many members expressing feelings of deep betrayal and hurt, such as a lister who had worked very hard putting together events for Seymour's "Starweek," when she was given her star on the Hollywood Walk of Fame the previous year:

> I hope that she did think about what she said afterward... I understand she was feeling hurt but surely she must remember it was these same "internet" fans who stood out there a year ago shouting "we love you Jane"... Does she truly believe now that those same people are suddenly "out to get her?"... I can't express the sadness I feel at the moment. It hurts to think that Jane would believe what she seems to about us after all that! I meant what I said about her last year when I made that speech. I still feel that... I know there are a lot of actresses that are more than satisfied with just "fans," but Jane has something very different and I can't believe she would want to lose that... She can have just "fans" who will walk without a backward glance if they don't like her latest project or she can keep the kind of "intimate fans"

that she has—the ones who will watch no matter what the reviews say because they care about HER. . . .

Another, who had also been deeply involved with various organized activities on Seymour's behalf, is less charitable:

> I feel that Jane doesn't get the right info from her fans because she really doesn't care enough to find out what is really going on . . . We are most often an after thought until we are of use, and it doesn't sound very nice but in many ways it is true with these stars . . . we are not fools here . . . when Jane needs her fans again, we will be those wonderful Internet fans once more. . . .

Alongside many expressions of anger, members begin to assert their identity as a community, united in their reaction to the woman who portrays the character around whom that very community formed:

> She has made herself appear ungrateful and ungracious in a very public forum, and she's angered many people who have long supported her career. This list is *our* community; she's not a member. She doesn't belong here until she joins DQMW-L. If anything less than unconditional positive regard is upsetting, then maybe she shouldn't have been listening at the door that someone deliberately left open?

Finally, a regular lister posts a very long commentary on the very notion of "fandom," comparing the relationship between fan and performer with that between friends:

> It would be easy for me to say I love everything she does, if I were a normal "fan." But to me . . . there are two [other] categories of fans. There are the fans who like her as an actress, could care less about her as a person, and will watch if they like what she does and leave if they don't. Then there is a different group, the one I believe I and many I know belong to. We are the "intimate stranger" fans. We are not "friends" of hers in a real sense, but we care about her as much as one can "care" about a "stranger." I think this kind of "fandom" and friendship really share a lot of the same elements . . . a good "friendship" must have three very important elements.

She goes on to discuss "honesty," "trust," and "support," suggesting that fans have the right to voice their true opinions, that stars should trust their fans' judgment, and that they should appreciate the support fans give, even if it includes criticism.

> In the case of the movie however, many felt this choice . . . would ultimately be hurtful for her career. So we voiced our feelings, most of us NICELY, I believe but yes, passionately as well . . . I hope she can understand that being a "good fan" may not always mean telling her we agree with everything she does but that we will continue to care and want the best for her!

Indeed, the relationship between performer and fan in the electronic age is a complex and volatile one. By 2002, the earlier hostility was no longer apparent as fans marked the 10th anniversary of the day filming began by organizing a reunion weekend, involving communal meals, a beach party, and organized tours of the national park where much filming was done, and the remains of the regular set. The climax was a dinner attended by many DQ cast members, including Seymour and Lando, bringing performers and fans together to acknowledge their mutual interdependence. The entire event is chronicled on the Web site for those who could not attend, with pages of cast/fan photos, and many personal accounts and tributes, including some from cast members themselves. Helene Udy, who played a minor but very popular character, reformed prostitute Myra, wrote:

> Without those of you who welcomed and built on that message and proved its importance . . . to the network . . . Dr. Quinn would have had no life. As I said in my little moment with the microphone, it is such a wonderful and continual shock to be in a room of like minds. To receive letters from people who value the importance of a show with a wonderful and hopeful message about human potential and human love and honor . . . I thank you all for being part of the circle of love, and making me a part of it as well. As Jane so aptly put in her moment on the floor, the word "fan" is not appropriate for the kind of people you have all shown yourselves to be. Your contribution . . . has always felt more like a communion of like souls, than anything else.

Strange claims are certainly made about "fans." On the one hand, they are obsessed lunatics. On the other, as depicted by enthusiastic proponents of the "resistant audience," they are practically guerilla fighters, as in Lewis's description: "By participating in fandom, fans . . . enter a domain of cultural activity of their own making which is, potentially, a source of empowerment in struggles against oppressive ideologies and the unsatisfactory circumstances of their lives" (1992, 3). Even with its increased feminist identity, I am not ready to say that DQMW-L functions as a fight against oppressive ideology, and in any case, many do not appear to consider their lives "unsatisfactory." In fact, it is clear that members' Internet fan activities have added a rich and rewarding dimension to their lives given them, among other things, new friendships, interests in history and literature, an opportunity to try creative writing, or even a recognition that they are more "feminist" than they realized. The list, rather than being a haven for paranoid loonies, is a community focused around fan identity, but functioning for its member as much more than a "fan club." As one member, an 18-year-old college freshman who has been a member for five years, wrote in 2002,

> Being a part of this list has been a particularly special experience for me because I've kind of grown up over the course of the years with all of you guys . . . My experiences as a fan range from Star Week to the infamous Barnes and Noble Jane encounter to being a co-author of a Memory Book for Jane. All have been moments that I wouldn't trade for the world . . . The Memory Book gave me the chance to stretch my creative skills and create a role for myself in the series . . . [and I was able] to meet the woman . . . who acted as a personal and professional role model.

As Hills (2002) points out, the online fan experience does not simply mirror the experience of being an "offline" fan. Rather, the medium allows for a level of self-reflection that makes the community itself a focus of its members' analysis, and creates an additional body of text that takes on a life of its own. The online community "perform(s) its fan audiencehood, knowing that other fans will act as a readership for speculations, observations, and commentaries" (Hills 2002, 177).

Much of the criticism of virtual communities focuses around the notion that they are replacing other kinds of interaction, and that "communities" that used to exist are threatened by disembodied Net interaction. I believe it is undeniable that virtual communities do not function in the same way as place-based communities. For one thing, virtual communities are "low-risk." In a virtual community, we are not forced to deal and interact with people we dislike. Although virtual groups do indeed have to work to maintain the community, any individuals who lose interest can easily drop out, and will soon not be missed. Members of DQMW-L who professed their devotion to the list two years ago have quietly disappeared, no doubt moving on to other "communities."

Yet cyber communities can also offer something that interpersonal ones cannot. Although much of the sense of community derives from a conscious attempt to replicate "real-world" connectedness, the nature of electronic communication adds an element that is quite distinctive. E-mail communication depends entirely on the written word, and so lacks the visual and oral cues that guide face-to-face communication. On a list, an individual's identity is constructed by verbal means alone. Clearly this allows for the construction of one or more completely different personalities, as happens in Multi-User Dimensions (MUDs) and other venues. Lists like DQMW-L stress the need for honesty and openness; people report that when they meet other members at the Pranch or other gatherings, they are usually delighted to find that their cyber "friends" are much as they expect them to be. However, they also write about how the facelessness of e-mail has opened up new communication possibilities. A teenage lister, for example, wrote about how she has developed close relationships with women who are much older than she, pointing out how this is unlikely to happen in "real life," as people make assumptions about others based on age, appearance, clothing, and so on. Others told me about their initial surprise upon realizing that another member was African-American; the unspoken point was that e-mail communication allows people to by-pass the often unconscious assumptions we make about race and other external "markers." Thus some barriers that function in real life to impede communication may disappear in the virtual community.

THE CHALLENGE OF INTERNET ETHNOGRAPHY

For the ethnographer, virtual communities are also low intensity, and low risk. To do my "fieldwork," I log on when I feel like it, and I can ignore the list messages for days at a time if I like (which is, of course, the way many participants also use the forum). In this context, participant observation is in a sense one-dimensional—all the fieldworker has is words and pictures. And the words that provide the raw data for analysis are written words, not spoken; they are crafted by people to communicate in a way different from everyday interpersonal communication. As one lister commented, after reading an early version of this chapter,

> looking back on those impassioned discussions seems somewhat overblown given the coolness of elapsed time . . . It's like the emotion and passion are stripped of their clothing and laid bare . . . It's like we can read it with our heads, but not always with our hearts— at least not to the extent the hearts were involved at the time.

Gajjala (2002) stresses the dangers of equating online and real life ethnographies, because of the crucial differences in interaction. She also studied an e-mail discussion list, and after meeting some of the members face to face, she began seeing them as "multidimensional human beings, something I was unable to do through reading their posts . . . " (p. 187). At the same time, she addresses the interesting potential that online ethnography opens for more genuinely collaborative feminist research: "The interactive nature of the medium potentially leads to a questioning of the researcher's conceptual and methodological assumptions by 'subjects'" (p. 182). I felt from the beginning of my study that, even though the list is in a real sense "public," it was not appropriate for me to lurk unannounced, and to quote members' words without their approval (see Bird and Barber 2002 for a more detailed discussion of the ethical issues raised in this study). My decision to announce my plans was important to me, as was the opportunity to share my analysis with the list. My posting of an early version of this chapter not only brought me valuable feedback, but also solidified my position on the list, since all the comments, whether written to the list or to me personally, agreed that I "got it right," while suggesting areas I might develop: "It captured the flavor, the intentions and the agonies of the list community so very well"; "it was both fascinating and deeply gratifying to read." I was fortunate; Gajjala's (2002) piece focuses on the consequences of a vote taken by "her" list that denied her the right to study and quote their interaction. If the DQ list had done the same thing, I, too, would have had no choice but to comply. In the same vein, I did not have to deal with what to do if members had vehemently disputed my claims. I even wondered if their approval showed that I had lost my objectivity and critical distance, and "gone native." But in effect, the ethical dilemmas of Internet ethnography are not drastically different from "real" ethnography—except perhaps that the temptation to be intrusive is actually greater, because one's invisibility makes it so easy.

CONCLUSION: FANDOM IN THE VIRTUAL COMMUNITY

In sum, I believe it would be short-sighted to dismiss virtual communities as hollow and insignificant just because they are not like "real-life" communities. Doheny-Farina argues that there is such a thing as a "true community," which must be bound in some way by geographical place and which "is a collective . . . in which the public and private lives of its members are moving toward interdependency regardless of the significant differences among those members" (p. 50). He contrasts them unfavorably with "lifestyle enclaves" (Bellah et al. 1996), usually based on connections formed by leisure or consumption, although one could also add work, religion, and any number of shared interests. These, he writes "are segmental because they describe only parts of their members' private lives . . . and celebrate the 'narcissism of similarity' through the common lifestyles of their members (p. 50)." Aden (1999) cites studies showing how solitary consumption of technology has so often replaced face-to-face interaction, and thus "contributes to our growing sense of displacement as members of a social community" (p. 35). However, I wonder how long it is since the organic, place-based community was actually the focus of most Americans' lives? Haven't we been moving for some time toward a culture based on interlocking networks rather than holistic community? This is especially true as women, who traditionally did much of the work of maintaining community, have moved out of the place-bound, domestic environment, and into the network-based larger world. In that sense, the Internet is not new, but simply one facet of an increasingly complex society.

There is indeed something almost frighteningly individualistic and narcissistic about certain activities on the Internet, and if anonymous, random interaction was largely *replacing* "community," maybe we should be concerned. However, DQMW-L shows that the need for real inter-personal connection is still strong, and that in certain contexts, Internet communities offer opportunities for that connection. Like the Internet, media generally are often characterized as driving wedges between people, as we all watch our televisions in solitude. Internet fan groups like DQMW-L show that the media *can* function to bring people together. One lister writes how she was a devoted, but secret fan before discovering the list: "I knew no one else who had this devotion . . . then I found this place, and I was no longer alone . . . I went further than I ever would have predicted, and my 'fanness' grew to encompass aspects that never would have happened if not for this List."

Our culture is permeated by mediated images and messages, and it is important to understand the implications of this, rather than merely bemoan them. The DQMW-L list, although not necessarily typical of the vast numbers of Internet-mediated interactions that occur every day, shows that many people are not passive, but strive to enrich their lives through media. Many members speak of their involvement in "real-life" communities too, as active participants in churches, schools, and organizations. We should not assume that virtual connections necessarily wipe out face-to-face ones.

Invoking a nostalgic notion of the past, critics often decry electronic communication as destroying community. It is perhaps ironic that DQMW-L members employ equally nostalgic visions to characterize their community. As Aden (1999) writes, one of the defining features of modernism is nostalgia: "When we don't quite know where we are or where we're going, we can return to a favorable past...We transcend time and cultural space by going back to the past while living in the present" (p. 43). Indeed, the DQMW cyber fans quite consciously use the most complex communication technology we have known to create a community based on a shared image of what an imaginary, yet historically specific, ideal community might have been.

4

IMAGINING INDIANS
Negotiating Identity in a Media World

INTRODUCTION: THE INDIAN AS A CULTURAL ICON

It's one of the most celebrated images of the American Indian.[1] A tight close-up of a middle-aged man, his stoic face speaking for the suffering of generations, watches the despoilment of his ancestral lands by the heedless pollution of the White man. The depth of his agony is revealed by the single tear forming in the corner of his eye. This 1972 public service advertisement, targeting environmental pollution, became an American popular icon, solidifying the environmentally-conscious, spiritual "noble savage" as the prevailing archetype of the Indian. The actor who posed for the shot, "Iron Eyes" Cody, seemed to personify that archetype in his professional and personal life.

More than twenty years later, when Cody died, his personal identity began to unravel, as it emerged that his real name was Espera (or Oscar) DeCorti, an Italian-American with no Indian ancestry. He had lived out his life as a kind of 'going native' fantasy, marrying an Indian woman, adopting Indian children, and acting as an Indian spokesman. He was only the latest in a line of Indian "wannabes" that included Englishman Archie Belaney ("Grey Owl"), and "Cherokee" Chief Long Lance, who was raised as a southern Black or "colored" man (Francis 1992). Indeed, many of the "Indians" who most caught the public imagination turned out to possess not a drop of Native blood.

Into this anecdote are packed many of the elements that make the role of the American Indian in popular culture so distinctive, not only in the United States, but also elsewhere. The original inhabitants of the

Americas were decimated in what amounted to genocide in the eigh-teenth and nineteenth centuries, and they are still one of the most dis-advantaged groups in the country, suffering from high levels of poverty, unemployment, and sickness.[2] At various times in the past, and even to-day, they have been stereotyped as savages, cannibals, sexual predators, and shiftless, drunken losers (Berkhofer 1979). Yet at the same time, they have been exalted as noble and spiritual—the true symbol of America, and the source of wisdom for a culture that has gone astray. White people have been "playing Indian" for a century or more, and they still do, meet-ing on weekends all over the United States and Europe, to dress in Indian regalia and act out Indian rituals (DeLoria 1999). Indian shamans, real and spurious, have become rich from teaching White people supposed Indian lore, and initiating them into rituals like sweat lodges (Whitt 1995). According to mainstream culture, "real" Indians are wise, calm, spiritual—and living in a kind of mythical nether-world. If a football team chose a grinning Black stereotype as its symbol, and called itself the "Coons," there would be uproar. Yet the Washington Redskins con-tinue to use their cartoonish mascot—after all, who is to be offended, since there aren't any real Indians around any more? (Davis 1993)

It's hardly surprising then, that images of Indians in contemporary popular culture are limited and one-dimensional, with a heavy over-lay of romanticism. There is now a substantial body of literature that documents the representation of American Indians in popular litera-ture, television, film, and so on, extending to representation in text books (e.g., Ashley and Jarratt-Ziemski 1999; Berkhofer 1979; Bird 1996, 1999; Churchill 1994; Francis 1992). American Indians are popular not only in the United States and Canada, but also in Europe, where Indian hobbyist groups are widespread (DeLoria 1999), and where novelists like Karl May in Germany established a tradition of "going-Indian" roman-ticism. Contemporary Indian life is rarely represented, and occasional recent films, such as the Indian-directed *Smoke Signals,* a funny and honest portrayal of modern reservation life, cannot compete with the long-established narrative resonance achieved with such blockbusters of noble savagery as *Dances with Wolves.*

Taken as a whole, this intertextual melange of imagery suggests that American Indians as symbols are a potent presence in North Ameri-can and European cultural narratives, while one wonders whether these narratives speak in any way to Indians themselves. As cultural stud-ies scholars have long argued, we cannot presume to read the cultural meaning of anything through textual analysis alone. Yet very few schol-ars have studied audience response to American Indian representation, from the standpoint of either Indian or non-Indian audiences. Hanson and Rouse (1987), having acknowledged the consistency of the noble savage imagery in contemporary culture, write, 'far less certain is the precise connection between the familiar caricature of the generic Indian and the more complex set of beliefs and attitudes that individuals actually had concerning Native Americans' (p. 57). My purpose in this study was to explore this connection, extending the discussion not only to how White audiences respond to representations of Indians, but also how Indians respond to, and imagine representations of themselves. Hanson and Rouse (1987), using a social scientific survey of students' attitudes toward Indians, conclude that their (mostly white) respondents' perceptions tend to be positive, rejecting the older stereotypes of war-like, primitive people, and claiming to value the contributions of diverse Indian cultures. They paint a picture of a rather open-minded population, who value diversity (a message they were probably used to receiving in the sociology and anthropology classes from which they were recruited). Their respondents rated television and movies as their most important source of information about Indians.

Yet Hanson and Rouse's data also point to the acceptance of stereotypes. For instance, the students associated Indians with rural and traditional lifestyles, and tended to think of them as living in the past. Respondents agreed that Indians tend to be 'submissive' and 'withdrawn,' even as they disagreed that Indians were 'lazy,' 'weak,' 'undependable,' or 'unpatriotic,' all of which might have described popular perceptions in the nineteenth century.

I believe this study shows that by the late twentieth century, college students knew how to give an appropriate response to a survey on racial attitudes, and few are likely to offer strongly pejorative comments. Indeed, it shows both the strengths of survey research—its ability to paint a broad picture using a large, representative sample—and its weaknesses— its inability to penetrate beneath surface attitudes. Although these students rejected the old Western stereotypes, they appear to have internalized the more contemporary stereotype, without seeing it in any way as derogatory. Prevailing American attitudes would not favor "submissive" and "withdrawn" as positive descriptions of most Americans, yet it seems unlikely that these students meant these terms as negative. Rather, they accepted the stoicism and laconicism of Indians as part of the "noble savage" paradigm. While Hanson and Rouse read their results as indicating a breakdown of stereotypes, I believe they merely suggest a shifting to a "positive" stereotype that reflects a self-conscious "political correctness."

Support for this view comes from my earlier research that compared Indian and White responses to *Dr. Quinn, Medicine Woman,* an Old West television series that featured Cheyenne characters (Bird 1996). This research had been based on focus groups with Indian and White viewers, who were asked to view episodes of the program privately, and then discuss them in an open-ended format. Group members were asked to discuss the Cheyenne characters in much the same general terms as other characters (that is, they were not asked questions about whether the representations were "positive"or "negative"). My conclusion was that the largely stereotypical presentation of Indians (stoic, non-emotional, spiritual, and so on) was accepted as authentic and essentially unremarkable by White audiences, while Indian viewers found it inauthentic, irritating, and one-dimensional. A typical contrast was in response to an episode in which the main Cheyenne character is unjustly imprisoned by the Army. He refuses to protest, even facing a mock firing squad without showing emotion. A White woman remarked approvingly on this: "Indians are like that. You know, they can be very intense emotionally but able to suppress it and not show it." (Bird, 253). Indian viewers were especially angered by this story, arguing that "his manhood was suppressed," and the character was not allowed to show normal emotions. "He just . . . put his head down, made him look pitiful. That kind of pissed me off" (Bird, 256). Similarly, Shiveley, in comparing male Indian and Anglo responses to a classic Western, *The Searchers,* found that while both groups enjoyed the film, the Anglo viewers also thought it was authentic in its portrayal of the Old West, while the Indian men rejected the dehumanizing they saw as central to the imagery, and did not perceive its historical representation as in any way authentic.

Thus, I suggest that while contemporary perceptions of American Indians are not generally "negative," in the sense of Indians being classified as savage, demonic, lazy, or drunk, they are still narrow, and ultimately objectifying. Hanson and Rouse's students may have known the right things to say about Indians and their cultural contributions, but their perceptions of Indians are still framed in a particular way, with media as central agents. As Thompson (1990) puts it, a central role of media is the "public circulation of symbolic forms" (p. 219), and the Indian is undoubtedly a potent symbolic construction in America and elsewhere. Many Americans, and almost all Europeans, have little direct contact with Indians, and so popular depictions are especially important. Indians themselves simply do not find symbolic representations that resonate in any way with their own experiences and identity.

ETHNOGRAPHY AND RESPONSE

Both my study and Shively's work were based on analysis of qualitative focus groups with Indians and non-Indian people, and both pointed to the importance of qualitative, open-ended methodologies in drawing a more subtle picture of audience response. Both studies fit within the tradition of "audience response" studies, focusing on a specific text, and they begin to get at the way that media representations connect with people's sense of identity, in a similar (though less extensive) way as studies like Bobo's (1995) on Black women reading the film *The Color Purple*. As defined by Woodward (1997), "Identity gives us an idea of who we are and of how we relate to others and to the world in which we live. Identity marks the ways in which we are the same as others who share that position, and the ways in which we are different from those who do not" (pp. 1–2). For instance, a central character in *Dr. Quinn* was Sully, a glamourous, long-haired loner who has lived with the Cheyenne and "knows their ways." White viewers, especially women, liked him, seeing in him an ideal hero: "He stands up for the women and... the blacks... and the American Indians and... he's always doing the right stuff," commented one (Bird, 255). Sully unproblematically fits with a White sense of identity related to the West and the Frontier, in which the role of the good White man is to relate to and speak for the oppressed, but ultimately to guide them gently toward the inevitable progress of civilization. He comes from a long line of border-crossing mountain men, traced back to early pioneer narratives and the novels of James Fenimore Cooper, and he became a staple in the Western narratives of Europe, such as the still-popular German novels of Karl May. He is the personification of the "going native" fantasy that fuels Indian hobbyists and role-players (Baird 1996). For all its mainstream resonance, this is a narrative that violates Indians' own sense of identity. They recognized the theme: "I can't think of one movie that there hasn't been this White guy that has somehow been part of their culture" (Bird, 255). But they resented it strongly: "Here's another White person fixing the Indians" (p. 255), "I know a lot of old stories... I can't ever recall one where anyone talked about a long-haired, light-skinned, hairy guy that helped my tribe" (p. 256).

Discussing the relationship between identity and media representa-tions, Woodward (1997) writes, "Representation as a cultural process establishes individual and collective identities, and symbolic systems provide possible answers to the questions: who am I?; what could I be?; who do I want to be?" (p. 14). For Whites and Indians, the experience of watching the *Dr. Quinn* Indians and wannabes, or "classic" movies like *The Searchers*, is diametrically opposed—Whites' identity is vali-dated and authenticated, while Indian identity is denied and erased. Indian participants in my study spoke about how media representation is so rooted in mainstream representations that it may be impossible for conventional media to provide the kind of narratives that speak to Indian identity. An Ojibwa participant spoke of the need to transcend conventional media forms and reinvent identity: "I think what's coming up now is virtual reality experience rather than just one dimension. We're going to have something more than T.V., where . . . our people will win the game" (Bird, 259).

CREATING A NEW MEDIA EXPERIENCE

The earlier studies suggest that Indian people do believe media representations are important, both for their sense of personal identity and as literal mediators through which relationships between themselves and people of other ethnicities are filtered. The daily experience of American Indians is that White people constantly see them through the lens provided by the media. Ojibwa writer Jim Northrup writes about traveling to New York City in the aftermath of *Dances with Wolves,* and finding himself treated as a minor celebrity and curiosity, because of his Indian appearance. "One cab driver took a $10 bill out of his wallet for us to autograph. I signed as Kevin Costner [the star of *Dances*], and I believe my wife used the name Pocahontas" (Northrup 1995). The film *Smoke Signals,* based on the writing of Sherman Alexie, a Coeur D'Alene Indian, and scripted by him, includes a sequence in which the two young central characters take a long bus trip from Washington State to the Southwest. The more confident Victor instructs his nerdy, bespectacled friend Thomas how to look like a "real Indian," exhorting him to let his long hair flow free, "look stoic," and "look like you've just come back from hunting a buffalo." Thomas protests that they are from a tribe of fishermen not hunters, but Victor knows that the popular image of Indians is a generic Plains Indian buffalo hunter: "This isn't *Dances with Salmon,* you know!" At a rest stop, Thomas attempts to change his image, with limited success. However, reality sets in when the pair realize their seats have been taken by two surly, mildly racist white men, and they must move or risk a confrontation. The message is clear—mainstream society loves the Indian in his proper, mythical place, but in real life, Indians are still second-class citizens.

BEYOND RECEPTION STUDY: MEDIATED INDIAN IDENTITIES

With these points in mind, I devised a study that I hoped would offer a more subtle and nuanced understanding both of White perceptions of Indians, and of how Indians might re-imagine the role of media in representing themselves. Working in Duluth, Minnesota, I recruited 10 groups, each with four participants (although 2 groups ended up with three, due to last-minute drop-outs). Two groups consisted of White women, two of White men, two of Indian men, and two of Indian women, while one comprised Indian and White women, and another Indian and White men. Since the task involved a group planning exercise with minimal direction and supervision, I decided it would be most effective if the members already knew each other; usually one individual was contacted, and he or she was then asked to recruit three friends/acquaintances. No formal attempt at randomness was made, although I wished to avoid a sample only of traditional-aged students. Some Indian participants were recruited through the city's American Indian Cultural Center, others through programs at the University of Minnesota, Duluth's School of Medicine, which actively recruits Indian students. The result was a pool of 38 participants, ranging in age from 17 to 58, with an average age of 30. Each participant was compensated with a check for $50.

The groups were told that their mission was to design a fictional television series they would want to watch regularly.[3] It could be any genre (sitcom, drama and so on), and could be set anywhere, at any time period. The program should include a cast of characters, both major and minor, and the group had to decide who they were and how they related to each other. Their goal was to design the program in as much detail as they could, providing a history of the characters, developing a detailed story for the first episode, and outlining some of the events and storylines that would happen over the next six weeks. At the end of the estimated two hours, they were to summarize their decisions on forms provided. The sessions were also audiotaped, and the transcripts analyzed. The only restriction was that at least one character should be White, one American Indian, and one a woman, although none of these had to be a leading character. I hoped to avoid focusing the group's attention too heavily on the issue of Indian characterization; the groups were simply told that the project was a creative experiment. After full instructions, coffee, and soft drinks were provided, I withdrew while the groups talked. My decision not to interact with the participants was deliberate; as far as possible I wished for this to be an "ethnographic encounter" among themselves, and not between myself and them, thus minimizing the possibility of my agenda becoming a central focus. In fact, my research assistant recruited and set up some of the groups completely independently, so that I never met some of the participants.

Clearly, this is not an "audience response study" in the familiar sense; there is no media text to view, and no viewers watching and decoding.

This choice emerges out of a growing dissatisfaction among audience researchers about the limitations of response studies, as discussed in Chapter 1. In a media-saturated culture, it is no longer possible to separate out the "effects" of particular media (if it ever was), and the goal must be to reach a more holistic, anthropological understanding of how people's world views are patterned by the media, and how the media are inserted into their daily lives. As Drotner (1994) argues, for media ethnographers this implies a move away from the specific reception studies: "In empirical terms the context of investigation is widened to include areas beyond the immediate situation of reception . . . . media ethnographers apply a variety of methods in order to better grasp the dynamics of mediated meaning-making . . . as part of everyday life" (p. 345).

CREATING THE SERIES

Most of the groups began in varying degrees of despair, discussing how impossible the task was, and wondering about the point of the exercise. One participant commented, "Did you see the Seinfeld, this is kinda funny, Seinfeld writing a pilot for his own show. That's what I kind of feel like right now." However, all eventually got down to the task at hand, and produced a variety of creative ideas (the $50 payment proved crucial, turning the exercise into a "job" that had to be done properly, rather than a frivolous activity!). Several groups began with drama, but almost all ended up creating comedies or "comedy-dramas," often discussing how humor can deal with serious issues and connect people who may not have much in common, a point echoing Drotner's (1994) observation that when she asked young people to create their own videos, "humour . . . was the genre everybody could agree upon" (p. 353).

In the end, each group transcript averaged over 40 pages, providing a wealth of ethnographic detail about how people actually integrate generic conventions, stereotypes, and their everyday experience in a mediated world. All groups talked at length about their need to have programming to which they related, and which spoke to them, and all drew to varying degrees on both their own experiences and their "media literacy"— their enculturated grasp of standard televisual generic forms, especially those they liked best. While there was much to be gleaned from the exercise, I will focus here on the implications of the American Indian characterization, and how talk about it helped shed light on internalized perceptions of Indian identity.

In analyzing and drawing conclusions from the data, I am aware of the need to remain cautious. Perhaps most crucially, there is the danger of essentializing the responses—of assuming that everything people say or do is because they are one ethnicity or another, rather than because of other aspects of their individual identities. We all inhabit many interlocking identities, connected with our gender, class, personal history, age, and any number of other factors. For instance, in the mixed male group an older man tended to set the agenda for the three younger participants; it was hard to determine if the rather hostile attitude of one of the other participants was because he felt marginalized as an Indian, because he resented the confidence of the older man—or may be he was just in a bad mood. As a researcher, I can listen to the tapes, pay attention to tone of voice, and study the words, but must always be aware of the numerous complexities of the social interaction I have set in motion, as well as the fact that this moment is embedded in a much broader cultural context. It was impossible, for instance, to find groups of people whose level of existing acquaintanceship was the same; some groups comprised people who already knew each other very well, others were acquainted mostly through such contexts as college classes, and in two groups at least some of the members were meeting for the first time. These factors clearly affected the nature of the conversations that developed, and in presenting my analysis, I have tried to incorporate this as far as I can. There are certainly times I would like to be able to retreat into the techniques of the social scientist, devising ways to control all the messy "variables" that interfere with certainty!

Nevertheless, it is well-established that, especially for people of color, ethnicity is indeed a dominating factor in self-identity, and certainly plays into their interaction with media representation. For White people, on the other hand, whiteness is essentially taken for granted, and accepted as the norm, while most of the Indian participants made it clear that for them, their ethnicity is something of which they are aware all the time, comprising one crucial lens through which they view the world, including the media. So while I am sure many factors played into the groups' interaction, there nevertheless emerged a consistency in the discourse about "Indianness," and it was that consistency that convinced me of the appropriateness of my conclusions.

WHITES REPRESENT INDIANS

In analyzing the White groups, I was interested in how their own creations would connect to earlier work, in which White audiences appeared to have naturalized familiar stereotypes, and did not appear very interested in the development of Indian characters. To a great extent, that turned out to be true. The first White group comprised four women: a 58-year-old office worker, a 35-year-old administrator, a 32-year-old university admissions worker, and a 42-year-old secretary. Although not all close friends, they all knew each other, and the group dynamic was lively and cordial, with all four participants actively contributing. Their favorite shows were all sitcoms: *Frasier, The Nanny, Friends,* and *Wings,* respectively. Significantly, they began by talking about their favorites, and formulating how "their" show would fit in this genre. Soon, they settled on the idea of a twenty-something husband-and-wife team (Scott and Jennifer) running a connected hairdressing salon and body shop (car repair garage). The show is called *Dents and Tangles.* Tired of big city life, they have moved home to Duluth to start their business. The action will focus around their struggles to make it, amid an array of supporting characters. They develop detailed character sketches of the characters. For instance, Jennifer is described in writing thus: "27 years old, BBA degree, cosmetology background. Excited to start her own business and establish herself as a manager. Her relationship with Glenn (Scott's dad) will be complicated, as she and Scott see themselves as equals profession-ally. She's very stylish." Seven characters are given this kind of detailed description, both in the group's written summaries and their taped con-versation.

At various moments, someone reminds the group that an Indian char-acter is needed. One suggests, "maybe an Indian girl could be one of the best stylists and have her be really flashy and beautiful," citing *Pocahontas* as an example. Later, the same woman mentions Marilyn, a character in the popular series *Northern Exposure,* set in Alaska and featuring several Native characters: "She was good, her heritage came through and she was really very quietly intelligent." Another agrees: "That was the first thing that clipped into my mind was *Northern Exposure.*" It was quickly de-cided that Marilyn could be exported from that series intact: "We could have her make some one-line saying at the end of the show . . . And in some episodes you could have her say something and leave the shop in the middle of the show, say something rather, um . . . prophetic, as though she knew what the outcome of the situation was going to be." The final written description of Marilyn read: "early 40s? (no one's sure). Goes to Jennifer once every six weeks for a quarter inch cut, never anything different. Very soft-spoken, wise, prophetic."

All in all, this group's result was revealing. The discussion of 'Marilyn' was cursory compared to the detailed development of other charac-ters, and the group relied entirely on an already-existing media char-acter, while all other character descriptions emerged through dialogue about personal experience and knowledge. *Northern Exposure*'s Marilyn was unusually well-developed (although still sketchy compared to the Whites; Taylor, 1996), yet the version imported to *Dents and Tangles* was one-dimensionally stereotypical—wise, spiritual, silent. In discussing the events that would ensue in the first few episodes, the group detailed the escapades of even the minor characters, except for Marilyn, as someone occasionally remembers and says something like, "Oh yes, then Marilyn comes in, says something wise, and all that." It's hardly a stretch to imagine professional TV programmers developing their characters in much the same way.

The second White female group consisted of a 30-year-old community organizer, a 36-year-old secretary, a 37-year-old community organizer, and a 38-year-old "homemaker/mom." All were regular television watchers, whose favorite programs were *Northern Exposure, ER,* and the female-oriented sitcoms *Grace under Fire* and *Roseanne.* They all knew each other quite well, and shared a perspective that supported progressive social ideas and community change. Their shared understanding and familiarity led to an easy rapport and an atmosphere in which everyone contributed. Reflecting their combined preferences, they developed a contemporary "comedy/drama"called *Mesabi North,* featuring the everyday lives of a mixed group of people in a Duluth apartment building. This group set out to make the situation "funky," featuring an eclectic mix of types. With much laughter, they seem to be gently satirizing the tendency in American shows to showcase a "diverse" cast. This immediately leads them into the need to have an Indian character, who at first will be the apartment owner/manager:

M: She's Native American, she's trans-gender . . . She's our main character right now.

S: She's a vegetarian. She's into Earth stuff, you know keeping the environment clean.

M: That's where that funky smell comes from.

S: She's got some funky smells coming from her apartment, incense you know,

G: Hippie stuff.

S: Crystals, all the crystals and all the rocks around her.

Soon they decide that this central character will be White, but "How about . . . there was an American Indian elder living in the building?" This is met with approval: "sort of like a mother, spiritual advisor?" The group goes on to describe a range of deliberately stereotyped characters—a single, Latina mother, a gay couple, a lesbian couple, a Hmong immigrant family, and so on. They assert that the humor will emerge from the interaction among them: "and then also a better understanding of different lifestyles . . . maybe," commented one. The end result is that virtually all the characters are stereotypical. Even given that, the group provides much more detail for some characters; the gay male couple is quite fully realized, with details of occupations, hobbies, and personal appearance, as the group refer often to their own gay friends. They have more trouble with the Indian woman, as they grope for appropriate descriptions:

S: Selma's our Native American Elder Women.

M: Yup, elder lady...

S: What's Selma's last name? She don't need a last name, we just go by Selma.

M: It could be like Selma Morning Star.

L: I like Selma Morning Star.

M: Make her last name sound more Native American... Black-feather... or...

S: Sunbear.

M: Blackhawk... or

S: Brownbear, Selma Brownbear.

M: Or Redbird.

L: Selma Blackbear. Or how 'bout Redbird?

M: She has beadwork classes and she sells her beadwork at the pow-wows in the summer.

G: Not just beadwork. She does it all.

S: S: Weaving and basketmaking and makes rugs...

In the plot outlines for the first few weeks, Selma gets little mention, after one episode in which other tenants think she is smoking pot (she is actually burning sage in a ritual). Once again, the group has drawn heavily on existing symbolic forms to create their character, and then cannot think what to do with her.

Both White male groups were slightly younger than the women. The first comprised a 25-year-old jeweler, who favored the sitcom *In Living Color*, a 25-year-old graphic artist, and a 23-year-old student, both of whose favorite program was *Friends*. The three were friends, who socialize together frequently, and their conversation is free-flowing and uninhibited. They name their program *Crazy Horse Casino*, setting it in a conservative Southern community, where a couple is trying to open a casino. At one point, the group gets into a long digression about the enormous (and presumably shady) profits made by Indian casinos. This group was probably the least engaged of any group, constantly having to return to task, while drinking whisky (not provided by me!). Their characterization is weak, and they are immediately flummoxed by the need to include an Indian character. Before they settle on the casino idea, they cast around for possibilities. One suggests a Western: "I mean,

how are we going to incorporate all three characters in something else?"
Certainly, if one were to look at the kinds of popular media texts that do
incorporate a White person, a woman, and an Indian, that would seem a
fair question. However, these young men agree that they are not familiar
enough with the genre, and move on to a sitcom. First they suggest a
White couple adopting an Indian child; their next attempt, drawing on
one of the member's experience in the jewelry business, is also strikingly
stereotypical:

> K: OK, how 'bout the setting is a jewelry store in like, downtown
> Duluth, and they have to deal with all the drunk Indians that
> come in.
> J: Hey . . . the one could be a recurring character.
> K: There you go.

However, they decide this is also going nowhere, and move on. After
several false starts, they decide on a male Indian, married to a White
woman, who opens a casino in the South. One asks whether an Indian
casino "would be too racially stereotyped," to which another replies,
"who cares?" and the third adds, "It's our show, dang right."

Drawing on more stereotypes that Indian casino profits probably derive from White expertise or Mob connections, one suggests: "How 'bout,
the thing that's really pissing him off is the wife runs it and he's the Indian
guy. So he's gotta stay home and take care of this little kid. Or he's like a
janitor or something." They go on to discuss the humor involved in having the Indian try to hide his ancestry while running an Indian casino,
settling on his name as Todd Crazy Horse. They continue to explore
various characteristics for Todd, having him sexually involved with several casino employees, while his wife fumes, and taking direction from
the Mob. They talk often about "dialog between Todd and Mary Beth
which would outline humorous conflict resulting from his heritage," but
cannot come up with specifics. In the end, it is clear they have no idea
how to develop the character, although they do a much more coherent
job with his wife and other supporting characters.

The second White male group included a 25-year-old cook, a 24-year-
old student, and a 27-year-old "unemployed graduate student," whose
favorite programs were *Seinfeld*, the fantasy drama *Hercules*, and *Friends*.
They did not know each other, but were mutual friends of my graduate assistant (a fact I did not learn until after the completion of the interviews).
They began by spending about 15 minutes getting to know each other by
discussing mutual friends, hobbies, jobs, and favorite TV shows, discovering many commonalities, and working amiably and collaboratively.

They quickly agreed on a sitcom, and discussed how they disliked the kind of planned diversity that characterizes many such shows:

J: I think they have over-killed that in a lot of the shows these days. Just decided that it must have this ethnic background, this kind of show...

H: I just don't like it when it is so obvious, because then it's like it's just the opposite. And then it's just like so fake then.

Between them they developed a rather rich scenario for their sit-com, *The Other Lebanon,* centered around a bar in a Colorado town, originally settled by nineteenth century Lebanese immigrants. Conflicts emerge between the locals and "rich, yuppie ski resort people," who are becoming increasingly numerous. This group was explicitly determined not to be stereotypical, and introduced some interesting characters, such as a highly-educated bartender and a female garage owner, both of whom are given a wealth of detail. The bartender, for instance, is described as "a kind of a helper with community counselor person. His character is not too witty... I don't want another *Cheers* or anything. He's caring, he listens, and he doesn't drink on the job or anything, but when he does drink he really lets loose and you get to see him do that once in while. That'll be in the third week... that'll be a real shocker." Later more details are added, such as "six years of college in philosophy" and "it could follow as far as him deciding in life what he wanted to do while he's a bartender. That could be more serious, you know, being in touch with himself..."

They tried hard with the Indian character, beginning with a false start in which they visualized a group of students as the main characters: "But um, there'd be a chaperone, and build off of the image of the wise American Indian, somebody that's kind of calm." This was rejected as clichéd, and the group settled on their character: "There's an Indian women, she's a musician, a good musician, she plays rock music, to traditional Irish, to Patsy Cline... She owns a profitable, kind of alternative movie theater... things you wouldn't see... She's not a shaman or anything." However, once established, the group found it difficult to know what to do with her, only mentioning her once again, to reiterate that she would defy expectations in terms of her musical repertoire. Meanwhile, their other characters and scenarios were described carefully and thoroughly, and the groups mapped out detailed events in four episodes, never mentioning the Indian woman after the first episode. The most striking thing about this group was that, even with their determination not to stereotype, they essentially found themselves unable to draw on cultural knowledge that would help them imagine a fully-developed Indian character.

INDIANS REPRESENT THEMSELVES

The White groups were clearly most comfortable with stories and characters that fit both their own lives and the mainstream media genres they experience every day, genres that take the White experience for granted. Minority viewers often have to read against the grain, and view representations through a lens that places ethnic identity in the forefront. Lind (1996), following Cohen (1991) and others, stresses the importance of "relevance" in audience interpretations of media—the fact that cultural identity is a crucial framing device through which people view media imagery. American Indians, like other minorities, spend their lives acutely aware of their ethnicity, and of media representations of it. In my *Dr. Quinn* study, for example, White focus groups rarely initiated discussion of Indian representation, taking it for granted. I usually had to raise it as a topic, something that never happened with the Indian groups, who always raised it almost immediately.

So it was not surprising that the groups in this study all took the opportunity to create a program that explicitly explored issues of Indian identity and life. The first male Indian group included four pre-medical students, aged 33, 20, 27, and 32. They listed their favorite shows as *ER*, the mostly Black sitcom *Fresh Prince of Bel Air*, *Seinfeld*, and *Star Trek: Next Generation*. The group members did not know each other especially well, having come into the Duluth program from Nebraska, Texas, Maryland, and Minnesota. Nevertheless, their rapport was strong, and their conversation was exceptionally fluent and good-humored, drawing especially on their common cultural experiences as minorities. Their comedy, *Red Earth,* evolves through a series of ideas that all involve people as outsiders, beginning with one participant describing his own life: "Coming from a reservation, becoming urbanized . . . and going back to the reservation . . . Being accepted in your traditional beliefs versus the societal beliefs . . . I actually wrote a story on my own a long time ago . . . I called it *The Glass Culture.*" Others then discuss having a White female doctor trying to be accepted on a reservation, or an Indian doctor trying to fit into suburban America. To this is added a story about an Indian student's struggle to become accepted as a doctor.

This prompts reflection from other group members, as they talk about the differences among minority experiences: "Even if you are a minority, the Black struggle is hard to relate to the Native American struggle . . . You got a Black . . . brought over here from another country. What is their stake in this? . . . But what can you tell a Native American? I have to be here, that's all. Struggle's totally different . . . nobody has a stake in this country but us." Initially, the group is concerned that they keep the program accessible, not alienating the mainstream by focusing only on

Indians, but gradually the ideas become more and more "theirs." The next scenario is a rewrite of history, in which the Indians fight back, forming a majority, with "casinos all over America ... Indians buy back all the land, and kick the intruders out." This iteration was called *Return of the Buffalo*—"one of the Indian prophecies, when the Buffalo get replenished the wars will start and then the Indians will take the land back." In another development, "It'd be real funny if some aliens came to the planet and said, 'all Native Americans stay here, and all Non-Native Americans get out.'"

> C: And then you'd have everybody in the world trying to find out who they married back in there. Yes, I'm 1/2,000ths. I need to be over here, and then you have a little line that says, "tribal papers" ....
>
> S: ... and you ask them what kind of Native American and they'll say Cherokee. Cause every White person I talk to says, "yeah, I got some Indian in me." What kind? "Oh, Cherokee."
>
> C: I mean you gotta have a show that Native Americans take over something ... even the apes took over ... God, *Planet of the Apes*. We can have *Planet of the Indians* ... We could have *Rain Man* ... Anytime he sees a White person he just pees on them.
>
> M: It's called *Red Earth*.

The idea of an Indian/alien takeover takes hold, with "buffalo being teleported down here ... Just like *Independence Day* ... all you see is a dark cloud." The mother ship, in the form of a giant teepee, invades Washington, D.C., using buffalo dung as bombs, and forcing a meeting with the White, female President ("so we take care of the woman and the Anglo in one"). Later they adapt the idea, positing a Native American planet which is exploring distant lands, the first being Earth. They send a scout (George, who takes his name off the Washington Monument) to Earth, and he has to adapt and make contact. No one believes his story, and he ends up in the jail, the "nuthouse," or the "drunk tank." Ideas flow thick and fast as his survival skills are tested:

> C: He spends his night on the town. He's running round these White people with hatchets ... You can see George Scout stripping the bones ...
>
> S: George goes driving.
>
> R. George searches for a mate.
>
> S: See George in Central Park roasting ...
>
> S: His ass ...

C: A human being on a stick . . .

M: Want some White meat?

R: Man, White people be complaining about this show. Scare the shit out of that grad student [*my research assistant*].

C: He'll say, "Is this what they really think about me? They look at me like I'm food!"

R: George starts selling drugs for extra money.

S: Starts selling drugs out of the Washington monument.

Eventually George meets, but eats the President, then goes home to Red Earth, and "everybody's pissed at him because he's changed." Although they now know "the Earth is an evil place," the invasion proceeds. Observes one participant: "suddenly it's become a dark satire."

The second Indian male group included two brothers—a 41-year-old video producer and social worker and a 46-year-old counselor—a 39-year-old unemployed man and a 36-year-old cook. Their favorite TV choices were sports, *Northern Exposure,* news, and the classic sitcom *M*A*S*H.* They all lived and worked in Duluth, and had known each other for some time, so their rapport was easy and the conversation fluent. They created a drama "with comedy" called *School Days,* set in the 1960s in an Oklahoma Indian boarding school.[4] Like the first group, their discussion gained an increasingly strong Indian identity as the time progressed. They began with a stereotyped scenario of a White couple who go to an Indian elder for enlightenment, reasoning that this was the kind of story that would appeal to the mainstream viewer. Gradually the seeker of truth evolves into Joe, an Indian man who has lost his sense of identity in a White world, and the elder is his grandfather, who counsels him. As this happens, we learn through flashbacks about his life growing up in the oppressive boarding school (something two of the participants had experienced). As one participant says, "this Indian . . . is being tormented by the matrons of the boarding school for trying to hold on to his ways . . . But he goes on to become well educated and graduates from the boarding school, maintaining his identity all the way through. No matter what they put him through. A lot of it's never told . . . That our people have success stories."

They envisage the young man visiting home, having been taught not to speak Ojibwa at the school: "He knows the language but he doesn't dare speak it . . . and then maybe he kinda hides it from his grandpa that he's being locked into a room or that he's being punished for singing or for praying or using the medicines." The series would humorously explore relationships among students as they defy the authorities, while also featuring compassionate teachers and friendships among students and staff. Remembering that a White character is required, the group creates Miss December, the school principal. "Everything she talks, everything she says is demeaning...she's a cold-hearted bitch." The group saw humour as crucial, and media as a key to reducing prejudice:

> Prejudice is based on...you don't have the knowledge...The thing we battle the most is stereotypes...No matter how successful we are, we're just an Indian or we're just a Black, or we're just a Mexican. So you turn that into humor...You turn that whole thing around and what you're doing while you're laughin', while you're learnin', is you're correcting stereotypes, and learnin' how to laugh at them, together.

The Indian female groups also tackled the issue of identity directly. The first group were all students, aged 17, 18, 19, and 21, whose favorite shows were *Fresh Prince,* (2), news, and *Seinfeld.* All were from different Minnesota towns, except one, a Navajo from New Mexico, and although they were acquaintances from college, they did not know each other particularly well. Nevertheless, the similarities in their ages, and their common experiences, led to a friendly, easy rapport in which everyone contributed. Opting immediately for a comedy, they began by laughing about a reservation setting "with everyone going around saying 'How!'" Tossing ideas around, they mention the White "wannabe"—"you know there's always one non-Native always just hangs around the Indians." They move on to visualize an Indian school, not unlike the previous group's, with a cast of Indian students and a mean, sadistic White teacher. Then one member suggests they "make it about something that you would never even think of an Indian being—how 'bout a car salesman?" The group runs with this, playing with stereotypes of Indian reservation life all the way. They name the series *Rez Rider,* a common term for the kind of beaten-up car found on reservations, deliberately contrasting that with the super-car that was the star of the old series *Knight Rider.* Their star is Melvin Two Hairs, an Indian "car dealer to the stars" in the exclusive Los Angeles suburb of Brentwood. He sells expensive foreign cars, and the humor revolves around the reservation lifestyle he maintains in the ritzy neighborhood. His "company car" is "Rusty," the Rez Rider, which is described as a character itself: "the typical rusted, taped-up, reservation ride." They create the car in rapid dialogue:

D: You know, you have maybe a flashlight for a head light...
M: ...and you can only get out on one side. And you have to open it from the outside.

L: You have a plastic window in the back...

T: ...Duct-taped seats, powder compact for rear view mirror...

D: You know, you have all your bumper stickers from the pow-wow just holding your fender up...

L: "Red power."

D: And you know, like his muffler is being held up by an old belt. Then the antenna is the clothes hanger. You ever seen that?

L: And you don't even have reverse. You gotta get out and push it.

D: A truck, one of those little bitty ones, that can only fit one person...somebody heavy, like totally hogs the seat. And everybody has to ride in the back, and it only gets to go up to 45 miles an hour...and oh...he's gotta have that one-eyed, three-legged dog sitting right there in the passenger seat.

Melvin is married to Ruthie, and they have "hellion" five-year-old twins, "like little rez kids who, when they see people they're sneaking up on them...and they're all dirty. All crusty." The requisite White character is Mrs. Dubois, a rich, sophisticated widow who keeps buying cars from Melvin, and being shocked by his reservation ways. One day Ruthie serves dinner:

D: And instead of having a whole, expensive dinner, it's fry bread, mutton and beans.

L: And some commod. orange juice [government commodities given to reservations].

T: Oh, and then they have that big block of butter. And the wrapper's still on it: "For Reservation."

Nevertheless, Mrs. Dubois is a classic "wannabe," constantly hoping Melvin will teach her about Indian lore and tradition. The message coming from the group is that Whites need a dose of reality when it comes to Indians—and the squalor of reservation life, even as comedy, is that reality, not the spiritual savage.

The second Indian female group included a 27-year-old social services secretary, a student who did not give her age, a 34-year-old human services case manager, and a 29-year-old educator. All were local, living and working in the Duluth area, and had known each other for some time. Their favorite programs were *ER*,(2), *Star Trek Deep Space Nine,* and *Star Trek Voyager,* and they began their discussion by talking about the need for more "wholesome" and "family-oriented" TV programming, particularly decrying "sex-obsessed" comedies like *Friends.* They decided to create an issues-based comedy-drama called *Migizi Way.* Like the other

Indian groups, they quickly decide to explore stereotypes, beginning with commentary on the sad state of Indian representation in TV:

> B: There are very few television shows that portray Native Americans as Native Americans. Like, *Northern Exposure* had some guy that was non-Indian portraying an Indian.
>
> I: Which one was that?
>
> B: Um . . . I can't think of what his name was on the show . . . He helped Maurice, and he worked at the store.
>
> M: Ed . . .
>
> B: That's it.
>
> M: He wasn't an Indian?
>
> B: He wasn't an Indian.
>
> I: Ed's not an Indian.
>
> B: He had to dye his hair black every week to keep it black, blond as . . . .
>
> I: Ohhh, I'm so disillusioned.
>
> B: He was Scandinavian, blond as they can come.
>
> I: I'm so disappointed, I didn't know that.

Briefly, they consider a show set "on the rez," but decide against it: "I don't think anyone wants to watch life on the rez. It's too close to reality. They don't want to watch drunken Indians stagger about the streets." Instead, they decide they "would like to see a show that did what *Cosby* did for Blacks . . . Have some role models . . . " They choose a present-day setting: "We don't need any more historical, romantic Indian pictures. We want reality here." Again, they debate whether "reality" is what they really want: "get the rez humor out there . . . that's funny. But just how much of the real life rez do you want to expose?" One points out that "reality" isn't all negative: "I mean Native Americans: some are doctors, some are nurses, and some . . . . I mean I know a doctor, a nurse that's Native American. So it's not like it can't be done."

Just as the White groups had to find a role for the "token Indian," this group debated the required White character:

> I: Yeah . . . how 'bout, eh, why does one have to be white? We can have a token white person.
>
> M: Let's have a drunk white person.
>
> B: No, no, that wouldn't be nice.
>
> I: How 'bout the redneck white person, comes in, that you'd have to deal with once in a while to educate him . . . that Jeff guy . . . Foxworth, that's it. [*Jeff Foxworthy, a popular "redneck" comedian*]

M: We'd have him come in . . .
I: Yeah, and he's not a malicious person.
M: He's just a total moron.
I: He doesn't know, that's all.

They go on to talk about the value of "guest-starring people like Native Americans that are stars, like Billy Mills . . . actually showing real life characters that are Native American, that are role models." They develop a promising scenario about an Indian clinic, drawing on their own experiences, saying it would be more "real" than shows like *ER*. They discuss the possible need for a strong male character, since most of the discussion has reflected their experiences as single mothers:

I: The other thing is, I have a hard time finding, looking for a strong male role model figure. Should we have him as a mythical figure? You know, as a shadow?
B: Kinda like the white buffalo.
I: A mythical figure . . . is there any? Is that a reality? . . . A strong male Native American. Do you know any?
T: No.
I: So we can make one up.
B: So it would be mythical.
I: It is mythical, right.
B: That's the saddest state of affairs.

Eventually, like the other groups' stories, their scenario grapples with Indian people's movement between two worlds. In their tale, Mary Migizi, a 35-year-old Indian woman, returns to a reservation in Minnesota, about to give birth to her fourth child. Migizi means "bald eagle" in Ojibwa; the group deliberately chose the combination of English and Ojibwa words for their character's name, after debating whether to name her Mary Eagle. Her husband has just been killed in a drunk-driving accident, and she had been making a living as a writer/illustrator in the White world. Now she returns to her life with her husband's family, and over the next few episodes, experiences a kind of rebirth herself, as she rediscovers her cultural identity. Episodes would feature Mary's struggles with bureaucracy; for instance, her husband was an enrolled tribal member, while she has to fight to prove to authorities that she is a "real" Indian. The token White character is Steve, a good looking publisher who is excruciatingly "politically correct"; he is "always asking if saying this would be offensive."

Indians and Whites Together

Only two groups included both Indian and White participants, and these each produced an interesting and rather different group dynamic. The male group included two Indian students, aged 19 and 20, whose favorite TV shows were *Friends, Seinfeld,* and "news"; and "cartoons," *Seinfeld,* and the Discovery Channel respectively. The White members were a 37-year-old student, who enjoyed *Seinfeld, ER,* and "news"; and a 23-year-old student who liked *Cheers, Seinfeld,* and *M*A*S*H.*

The group began amiably, with one member suggesting that the shared appeal of *Seinfeld* might make a good starting point, and agreeing that humor was the key. It was not long, however, before a little tension emerged. Most of the first few minutes is a dialogue between the two White men; the older man (T) then asks one of the Indian students, "What are you thinking, D?" He replies: "Yeah, comedy...but maybe something a little bit, kinda realistic." T replies that he only watches comedies; "I don't watch things like *NYPD Blue.*" D counters with "Why do you watch comedies?" as the other Indian (W) mentions that he does not watch "that stuff" (serious drama) either. D, addressing T in a rather challenging way, asks, "Are those even realistic?" W seems to be trying to defuse the situation: "Ah, I'd imagine some are. They're different from our culture." D continues to address T: "Do you like to see your blood, guts, sex, all that? Do you like that?" The younger White man, J, makes a joke about trying to work all that into a comedy, and the group gets down to that task, again with the two White men doing most of the talking, and suggesting various scenarios for an ensemble show. Tension emerges again as the issue of ethnicity arises:

T(White):	We could have a White woman, and an American Indian....um...a White woman and an American Indian and an African-American and...
W(Indian):	Why?
T:	You know, just to add some diversity, and then we can just add the whole array of ethnicities, like a Hispanic landlord, or something like that.
W:	Or Asian.
T:	Yeah, or Asian, we could bring them all into this.
D (Indian, *sarcastically*):	Or maybe you could have a blind guy who doesn't know the difference between colors.

The two White men try to ignore the sarcasm, but the conversation struggles on uncomfortably as they try to address the issue of ethnicity:

J (White):   That'd be a good idea. Because that would empha-size the, kind of, almost like a . . . you can tell their ethnicity without even seeing them or something, based on their behavior, or at least, that's where the jokes would lie. Not to make fun of, but to, eh, I guess make fun of it, it's a comedy.

T (White):   Yeah. Or we could do, eh, we could have a com-edy, you know, like you were talking about, poking fun at some cultural stereotypes or something. You know. Unless that's a real sensitive issue to anybody here.

D (Indian):   Or poking fun at the White majority population. That'd be good.

T (White):   Yeah. And you know, like um . . . you know, I think a lot, you know, is that kind of a sensitive, is that a sensitive area, what eh, um, do you guys have . . . Are there, what are like some of the stereotypes you feel about White people that we can poke fun at?

D (Indian):   I don't know.

J (White):   It doesn't have to be all based on that but . . . you could just look at normal things that, uh, I guess normal things . . .

T (White):   That people make fun of each other.

J (White):   Than to slant them each way.

D (Indian):   Just to be humorous.

T (White):   Unless that's, you know . . . That's one area we can poke humor at. You know, we can switch gears com-pletely and focus on something else, you know . . . that's kinda on a cutting edge type thing, I mean, and it could be kinda sensitive. Um, you know, we can switch gears completely and do some other type of humor.

J (White):   I think the best humor is from poking fun at characters . . . like *Seinfeld,* you poke fun at Kramer because of what he does or George because he's such a weasel . . . But sometimes these ethnic things get in there a little bit, it's not that major.

D (Indian):   Soup Nazi and . . . That's funny. The irony is funny, too.

The halting, cautious tone of the dialogue, peppered with frequent hesitations, is markedly different from the free-flowing discourse of the homogeneous groups. While it is apparent that in particular the personalities of D and T do not mesh well, it's also clear that their different worldviews are closely linked to their sense of ethnic identity. Left to themselves, I suspect that T and J, the two white men, would move into the kind of comfortable interplay that other White groups demonstrated, probably articulating the taken-for-granted whiteness of the other groups. The presence of the two Indian men prevents this, and seems to frustrate them. In turn, the two Indian men seem to want to address their sense of disempowerment, as the other Indian groups did, but express their resistance for the most part in sarcastic, conversation-derailing comments (D) or silence (W). Significantly, of the 13,900 words included in the entire session's transcript, 9,500 are spoken by T and J, while the two Indian men account for 4,400, a ratio of over 2:1.

The topic of ethnicity surfaces a few more times, as the group grapples with the need to include a "quota," as D puts it. Again, we see some irritation, as T tries to get them settled on the ethnicities of the "two guys and a girl" who will be main characters. There is some sentiment for the location being Colorado:

T (White):   Let's say, eh, guys that, from all over, that wound up converging for whatever reason, I mean, then they have their apartment there. How's that sound?

D (Indian): Like White trash, trailer park type.

T (White):   Is that what you want? You want that?... Well, be sensitive here, don't go, try not to be getting too offensive... just make it eh, you know a character, you know, just a character... we'll let the ethnicities... be just something aside. I mean... the show is the primary focus, the ethnicities are just coincidental, you know, and it doesn't have to be a White trash trailer park or anything like that, cause I think, you know that carries some eh, you know, that's getting kind of offensive. Yeah, baggage and what not...

W (Indian): I get you.

T (White):   So....

D (Indian): I thought it was humorous.

T and J try to lead a discussion of the characters. Unlike any other group, it is suggested that each individual takes ownership of a character, "playing" them, in a way. T presses D on "his" character, an Indian male D named "Wookie," trying to get him to flesh out the character:

D: I guess he came from around here and he's out there soul searching.

T: OK, so he's from Duluth originally, you getting this J? . . . How old is he?

D: Eh, 22, 23.

T: OK, how long has he been out in Colorado?

D: Not too long.

T: A year, two years?

D: OK, right.

T: Um, is there anything . . . about this character that's kind of . . . humorous?

D: I don't know. Is there?

T: It's your character, I don't know . . . So, you want to think about it for a while, OK?

D: Sure.

They move on to J's character, Sonja, a Scandinavian girl. J begins describing her as a typical blonde, blue-eyed beauty, but then defensively checks himself:

J: . . . But I mean, good looking in the Norwegian hair, um, blue eyes, and that doesn't always constitute someone's good looking.

D: But it does for you.

J: No, actually not, it's not a given by any means, by me.

D: You like fat sows?

The discussion limps on, as various members of the group tackle one character or another. Everyone studiously avoids mentioning ethnicity as any kind of defining characteristic; it is simply agreed that two characters will be Indian males. J voices frustration at how unsatisfactory all the characters seem to be, provoking comments from the two Indian men:

D: The essence of a person, I mean, how do you capture it on TV?

W: Yeah, that's what I'm saying. How could you do it unless you get real characters?

T: Well, like, so you want to add more to your character?

D: Well, what else, what do we do? Maybe to everybody, how do we, how do you capture a person?

W: How do you make it seem like they really do exist?

Everyone seems to be aware that something is not quite working. Finally, everyone responds to D's deliberately off-the-wall suggestion that Mookie is a tattoo and body-piercing artist, who pierces Gil (the other

Indian man's) scrotum, and causes infection. At last this is something all the men can relate to, and they engage in a long, animated debate about what would happen to a man's "sac" following this mishap, especially when his parents were coming to visit. This is followed by more discussion about the various characters, and the creation of vignettes involving Sonia as a veterinarian specializing in disturbed animals, or Stan as a frustrated chef. By the end of the session, the group has a set of essentially disconnected characters and some humorous, if also disconnected incidents, and the four part on apparently friendly terms. The lesson they all seemed to learn was that questions of ethnic identity are best avoided, in the interests of cordial relationships.

The female group was more successful in producing a coherent setting, plot-line, and cast of characters, although their show was not as fully-realized as most of the homogeneous groups. Their discussion was dramatically more cordial and collaborative than the men's. The group consisted of two Indian students, ages 19 and 23, from North Dakota and New Mexico respectively, both in Duluth for the College of Medicine summer program. The White participants were also students, aged 37 and 24, and not in the medical program. As a group, they did not know each other well before the meeting; in fact this was one of the two groups in which some members were meeting for the first time, having been brought together by mutual acquaintances. The group spent quite some time (almost 15 minutes) on "getting acquainted" conversation, discussing their favorite TV shows, and the qualities they look for in them. Finally, V, one of the two Indian women, suggests that they go around the table and introduce the type of show each would like to see. Comedies are suggested; when V speaks, she introduces the idea of a more "realistic" program:

> V: I guess this is my dream, maybe something about maybe how people see each other by what color they are rather than the inside. I mean, that goes from like if they're big on the outside . . . and most of the nicer people are big, and people just judge people on that . . . I know I would want—maybe just because of my background—I know I would . . . watch.

This idea is taken up with enthusiasm, and it is quickly agreed that the setting will be a college campus, and that the focus will be on how minorities find acceptance hard. Unlike the male group, the women take on the issue head-on, and seem to agree to learn from each other:

> P (White): Talking about Duluth, I was saying how I thought all the people were friendly and real nice . . . and this friend of mine I was talking to, um, was a minority

and she looked at me really funny and I said, "Oh, isn't that how you've experienced Duluth," and she said, "no, not at all, the people are really, um, covertly prejudiced, you know." They don't necessarily come right out and say anything, but it's the underlying way they treat you and everything.

V (Indian): I agree.

P: I've heard a lot of people say that about Minnesota in general.

V: I heard there was a lot of nice people... but um, coming from New Mexico where there's a large Hispanic and Native American population it was just so different. I mean, to this day um, like when we go to the mall, we just don't get service, we just don't. People just don't even say hi. That's a real big pain.

P: Up here or in New Mexico?

V: Up here.

D (White): Really?

S (Indian): Absolutely... I came originally from a reservation in North Dakota, and I've been around, I mean I'm attending college in Minnesota—Concordia—and I've experienced it all, with just no service. I mean, here I am, I smile and try to um, push away those stereotypes, I do my best to inform people that don't know about our culture... I mean, it's out there.

P: I dated some minorities when I was younger and just the things that my own family members would say to me... or people would stare at you sometimes... I just think that it's really awful for people to judge people... So that would be... do you guys want to do that?

S: Sure.

All four women began a spirited and friendly discussion of their own experiences of stereotyping and discrimination, making clear attempts to relate to each other, such as in P's story of leaving the blonde, blue-eyed world of Scandinavian Minnesota:

P (White): I remember one time I went to New York, and a lot of people have dark hair and so I like really stuck out like a thumb. It was just kinda different, because you're like used to being in an area where you kinda blend in, and all of a sudden... It didn't bother me... I just noticed it, because some of my friends said something to me.

S (Indian): Would it have bothered you, I don't mean to be . . . I'm just saying hypothetically, what if somebody did say to you, "you have to have these," would it have occurred to you, I'm away from home, I'm in a different environment . . . What would you have done if they called you, "hey you blondie, go back to your own . . ."

P: . . . like we talked about in this class, how impossible it would be to really put yourself in the other person's shoes because you just don't realize how much White privilege there is everywhere?

They talk about humor, and its potential to hurt, in detailed, personal stories:

S: I went to summer school to finish my first year . . . I was lonely and I was away from home and things and I'd get my mail and I overheard this conversation. This girl was taking a summer school class, they were studying Indians . . . I overheard it . . . um, they were literally mocking us, and it hurt me, it hurt me. As it is I was homesick . . . and they were saying things like, "What's the first thing you think of when you think of an Indian?" And he was saying savage Indian and hysterically laughing. And I was thinking . . . I have to be strong . . . that made me think that people literally think things are funny, but they don't think about it. Later . . . I approached them, and said how it had hurt me and things, and they literally switched around the story and they said they didn't mean any disrespect . . .

P: Good for you.

Every so often, the group works to apply these personal anecdotes to their TV scenario, as in an exchange about the sensitive issue of roommate allocation:

D: We could do a show, just on that, or on the roommate thing.

P: So who do we want for characters?

D: Well, we could have two women, I mean, one a Native American and one a White woman, and being roommates and just kind of trying to understand each other and deal with first year of college. That takes care of our characters that we need.

V: Or we could start the show with a student orientation and then you know, people are checking in with the dorm or whatever.

> And then go from there. Walking around, looking for a room or something.
>
> P: And walking past a room and then maybe a minority in the room and thinking oh, my gosh.
> S: And thinking, I hope that's not my roommate.
> D: So, those are the thoughts running through the person's head. OK.

They discuss in detail the kind of town the college should be in, and raise various general scenarios that might be addressed, such as dating, interaction with professors, freshman orientation, and so on. Naming the Indian character provokes discussion, with suggestions that it might be a name that doesn't sound obviously Indian:

> P (Indian): Maybe like Lisa or...Lisa Johnson, Thompson, or...
> S (Indian): My friend's name is May, but her full name is Maymingwa.
> P (White): OK, Native American female's name. How do you spell that?
> S (Indian): I don't even know how to spell it. Means butterfly.
> D (White): Oh, it does?
> P (White): That's pretty.

The group discusses which reservation May comes from, and what her academic and professional goals are, before moving on to her White roommate:

> V (Indian): Ok, her name is Sara Smith.
> S (Indian): Geeze, I don't know.
> P (White): What's her background?
> V (Indian): Are we going to make her snooty or...really wealthy?
> S (Indian): I don't know, I don't want to be stereotyping...
> P (White): Would say successful...successful family who owns a local business in town.
> D (White): Sure, there you go...So her name is well known.

More details are provided for both Sara and May, who will butt heads and have preconceptions about each other, but will eventually learn to value their friendship. At one point, they briefly discuss how to identify the two characters:

> V: Do you prefer to be called Anglo or Caucasian, or it doesn't matter? White?
> D: Doesn't matter...I never thought of myself as anything really.

P: What do you guys prefer?
V: Personally it doesn't matter. I mean, Indian is fine, Native American or American Indian.
S: As long as they have respect. The main word.

This group was considerably more successful than the male mixed group. They produced a detailed setting on the college campus, calling the program *Rainbow Ties*. Their two main characters were well-conceived, and both drew on the Indian and White members' personal experiences. Potential plots were rather weak and briefly described: "Then there'd be things about dating . . . " Most of their conversation consisted of personal narratives, which in themselves were very revealing about their experiences with ethnic diversity, and the kind of stereotypes that emerge in everyday life. While both mixed groups did not know their fellow members well, they each addressed this differently. The male group seemed consciously to avoid the obvious ethnic differences, producing stilted and sometimes hostile debate. The female group embraced the difference from the beginning, and through their discussion, they worked hard to see each other's point of view, exhibiting the collaborative style so often noted in female interaction (see chapter 3). They used the final 20 minutes or so to learn more about each other, not in terms of ethnicity, but rather in terms of personal history, career plans and so on, and parted on very friendly terms. Their TV program was less developed than most of the homogeneous groups, but that may be largely because of the need to establish a personal relationship; those groups that already knew each other well were exactly those who "got down to business" the fastest. For me, the very fact that we could not put together inter-ethnic groups of close friends was in itself revealing and a little discouraging.

CONCLUSION: NEGOTIATING IDENTITY IN A MEDIA WORLD

We know that television does not mirror reality (nor do people want it to), but that it refracts back a sense of reality that speaks to people in different ways. *Seinfeld* became the most popular TV show in America, not because it literally resembled the lives of its White, urban fans, but because it created a sense of recognition among them. *Seinfeld* never developed an audience among African-Americans, and it did not succeed in Britain, either. And for the mainstream American majority, there are countless opportunities to recognize themselves, and the genres that tell the familiar tales. As Livingstone (1999) writes, "Media cultures provide not only interpretive frameworks, but also sources of pleasure and resources for identity-formation which ensure that individuals . . . have a complex identity of which part includes their participatory relations with particular media forms" (p. 100).

The program scenarios created by the White participants in this study told tales that with varying degrees of creativity, felt quite familiar, reflecting their own taste for mainstream sitcoms. The main players were people like them, and they pulled from a wide cultural repertoire in which people like them might interact. They had difficulty placing their required Indian characters in their scenarios, and they were largely unable to develop their personalities. They drew heavily on media-generated stereotypes of nobility and stoicism; even when they consciously attempted to subvert those stereotypes, they did not know how to "write" the characters. This is particularly striking when one remembers that northern Minnesota, where the study was done, has a fairly significant Indian population, unlike many regions in the United States. As the Indian groups' discussion indicated (and in fact as the current occupations and career plans of the Indian participants showed), Indian people are to be found everywhere in the region, in any number of settings, yet to many White people they are apparently invisible.

For the Indian participants, the media world is clearly one they must negotiate. Although some did like popular sitcoms, the range of their favorite TV programs was quite different, with news and sport getting mentions, and appreciation for fish-out-of-water sitcoms like *Fresh Prince*. Several appreciated the *Star Trek* series, in which cultures often collide. Their scenarios spoke vividly about how it feels to be an outsider, and they revel in the chance for Indians to be the stars, and the winners. Their White characters, although often unsympathetic, and certainly at times stereotypical, are drawn from the personal experience of living as an Indian in a White world, and not so much from media images. The same can be said for their Indian main players—they are acutely aware of the prevailing Indian media stereotypes, and reject them angrily, in a way that speaks volumes about being marginalized in a world of alien media imagery.

I conceived this study as an attempt to extend the study of media reception in creative directions, striving for what Ang (1996) calls "an ethnographic *mode of understanding*" (p. 72, italics in original). One way to do this is to move away from confronting specific texts, finding ways to "back into" the question, as I have tried to do here. This study was not about how specific "audiences" respond to specific images, but rather was an attempt to explore how people construct their notions of reality by using imaginative tools that are largely given to them through mediated images. If White groups were largely unable to imagine Indians in non-stereotypical ways, it is not simply because they have watched *Dances with Wolves*. It is more because their cultural tool-kit contains only a limited array of possibilities, which have worked together over time and across media to produce a recognizable cultural "script" about Indians. *Dances with Wolves* contributes to that script; so do *Dr. Quinn,* Indian dolls, toy tomahawks—you name it. For most White Americans, to live in a media world is to live with a smorgasbord of images that reflect back themselves, and offer pleasurable tools for identity formation. American Indians, like many other minorities, do not see themselves, except as expressed through a cultural script they do not recognize, and which they reject with both humor and anger.

5

A POPULAR AESTHETIC?
Exploring Taste through Viewer Ethnography

INTRODUCTION: CULTURAL STUDIES
AND AESTHETICS

There was once a time when academics and cultural critics had no trouble making judgments about popular media. From the Frankfurt school's indictment of the homogenizing "culture industry" to MacDonald's 1957 dismissal of the "tepid, flaccid Middlebrow Culture that threatens to engulf everything in its spreading ooze" (MacDonald 1998, 27), the message was that "if it's popular, it must be bad."

Of course there are critics who still agree, including academics like Postman (1985), who worries that we are "amusing ourselves to death," and Twitchell (1992), who laments the decline of American culture into the vulgar and the carnivalesque. Indeed, as we watch the flow of TV images of personal degradation, from *Jerry Springer* to *Fear Factor*, it is easy to share that despair.

Yet in recent years, the voices of doom in media studies have been all but drowned out by cultural studies scholars, who have offered a major reappraisal of the popular, which Frith characterizes, rather tongue-in-cheek, as being "If it's popular, it must be good!" (1991, 104). In actuality, however, what has happened in cultural studies has not been a new aesthetic valuation of popular culture as "good"—a kind of mirror image of the "spreading ooze" judgment. Rather, the reappraisal has been essentially the replacement of aesthetic standards by political and social ones. More important than "quality" is the meaning of popular practices within specific social and political conditions, with a growing emphasis on the power of the media audience to make these meanings, and resist dominant ideological messages. It is not necessary for me to rehearse in detail the very good reasons for this, which emerged from an unarguable understanding that cultural distinctions are historically and socially constructed (Bourdieu 1984). In spite of the fact that some philosophers still argue for a conventionally formalist interpretation of the aesthetic, most critics acknowledge the difficulty of trying to create meaningful criteria for what is objectively "good" or "bad." Even Twitchell, whose entire book is a tirade against vulgarity, acknowledges that "Of course, aesthetic evaluation is partially social oppression, a refusal to share" (p. 18).

Raymond Williams's early appropriation of an anthropological definition of culture had challenged the established distinctions between high and low culture, and good and bad taste. As he wrote, "I find it very difficult... To identify the process of human perfection with the discovery of 'absolute' values" (Williams 1961, p. 57). Thus as we are analyzing a work of art, we need to consider its context in a particular time, place, socio-political structure, and so on; and we need to examine the ideological messages embedded in it. The issue becomes not whether an object or text is good or bad, but how it functions within a society. This understanding opened the door to the extensive reappraisal of popular media that followed in the last forty years, as the emphasis switched from issues of taste to issues of social context, oppression, and ideology. Within that reappraisal, an appreciation of the audience moved to center stage as, inspired by theorists like de Certeau (1984), scholar after scholar found evidence of "resistance" in readings of popular artifacts, with audiences in the role of "guerilla fighters making their raids on places of power" (Moores 1993). The trend was acidly critiqued by Morris (1988) with her widely-quoted comment that "I get the feeling that somewhere in some English publisher's vault there is a master-disk from which thousands of versions of the same article about pleasure, resistance, and the politics of consumption are being run off with minor variations" (p. 5).

To some extent the pendulum has swung back, and the most ardent advocates for audience power, such as Fiske (1989a; 1989b) have come in for fierce criticism, centered around questions such as the validity of the evidence for audience resistance (see Moores 1993 for a summary of the objections), or whether valorization of audiences may simply mask the subordinate position they are really in vis à vis media texts (as I suggest in Bird 1992a). I will return to a more thorough discussion of these significant issues in my final chapter; at this point I would like to take the discussion in a different direction—to consider how cultural studies of audiences might allow us to return, however cautiously, to the question of aesthetic valuation of popular media.

In the enthusiastic embrace of the cultural approach, what often seems to be forgotten is that Williams, and other founders of the cultural studies movement, worried about the implications of this new understanding of culture as a value-free term. They saw clearly that it raised crucial, and possibly unanswerable questions about taste and aesthetic value. After all, the cultural studies movement in Britain originated in part as one element in a workers' education movement, by which opportunities to participate in supposed "high culture" could be extended to those who had been unable to benefit from the elitist British educational system. As a working-class student who had defied the odds, Williams saw hopes for democracy in the idea of everyone having access to "the best that has been thought and said in the world," as Arnold famously put it. Along with his emphasis on social and political context, he argued that when we analyze a work of art, we must also speak of its transcendent qualities, its ability to address universal human concerns—and some texts seem to do that better than others.

Similarly, cultural studies guru Stuart Hall, writing with Whannel in 1964, was excited about the opening up of scholarship that made phenomena such as youth culture and popular music legitimate objects of study, not as examples of vulgar taste, but as cultural practices. They do sound a little tentative: "It might be said, then, that the pops cannot be judged at all – but have rather to be seen as part of a whole subculture, and handled as one would the chants and ceremonies of a primitive tribe. Are these standards anthropological?" (1964, 295–6). Yes, they conclude, that really is the right way to look at pop music. Yet at the same time, they are uneasy about jettisoning standards of quality altogether. "It invites a slack relativism, whereby pop music of any kind is excused because it plays a functional role in the teenage world" (p. 296). Finally, they argue, we cannot abdicate a responsibility to make judgment: "It is a genuine widening of sensibility and emotional range which we should be working for—an extension of taste which might lead to an extension of pleasure" (pp. 311–312). But, like those who followed, Hall and Whannel were unwilling to offer criteria as to what constitutes standards of "sensibility and taste" in popular culture, and few if any of their followers have ventured into that minefield since. Instead, as cultural studies gathered momentum, earlier misgivings about taste began to drop away, as the "anthropological standards" mentioned by Hall and Whannel became the norm. Anthropologists had long been wary of questions of taste in their studies of art forms in other cultures, arguing that value judgments would be ethnocentric, a truth that is indisputable (Banks 1998). Cultural studies scholarship followed the same model; guided by the relativistic approaches of anthropology, media scholars have managed to ignore or skirt issues of taste and aesthetic judgment in three main ways:

First, they frequently choose to study genres which they enjoy and value as fans, whether they take a purely textual approach, or incorporate an "audience study" component. This partly explains the now large body of literature about certain genres of rock music, written by a couple of generations of academics who grew up on it, and who accept its value (often inspired by such influential critics as Grossberg 1984a, 1984b). This trend has also led to a growing body of interesting self-reflexive work, that interrogates the relationship of the academic fan with other fans, while perhaps leaving unexplored the tendency described by Hills (2002) as "reserving for themselves the ability to determine political significance" (p. 13). For the self-defined scholar-fan, the issue of whether the fan object is "good" is taken for granted. Simultaneously, there is a usually unspoken understanding that other kinds of popular forms are beyond the pale. There is very little cultural studies work on music that does not suit the tastes of the scholar, such as the middle-of-the-road popular and country music enjoyed by vast numbers of people; we are much more likely to read a cultural studies article about the Grateful Dead or Phish than about Garth Brooks or Celine Dion. The middlebrow consumer, "the easy listener and light reader and Andrew Lloyd Webber fan," as Frith (1991, 104) puts it, is the most ignored, just as, with its values "more traditional than transgressive, country music has none of rock's radical chic" (Shusterman 1999, 222).

The same applies to television, where there is also a tendency to favor programs and genres that may be considered edgy, avant garde, or attracting a "cult" audience, such as the *Star Trek* franchise. I recall, for example, an International Communication Association meeting in which two whole sessions were devoted to the David Lynch TV show, *Twin Peaks*, a critical but not popular success. I have rarely heard a presentation about successful "middle-of-the-road" offerings like *Home Improvement* or *Diagnosis Murder*—and never from scholars who identified as fans. In fact, critical response to the success of such shows is generally wonderment about how such unchallenging programming can keep so many people in their recliners, week after week.

Kellner (1999) is more overtly contemptuous of the middlebrow than most scholars, arguing that television essentially comprises "neatly packaged narratives produced for easy consumption by teams of creators who aimed at making their products attractive and accessible to the lowest common denominator audience . . . following pat formulas, relying on generic conventions, and reproducing dominant ideas, and devoid of controversy or complexity" (p. 161). He then goes on to extol *The X Files* as a rare exception that employs a "post-modern aesthetic" to transcend and subvert the blandness of other "aesthetically impoverished" offerings (p. 163). For the most part, however, scholars do not care to define what is "wrong" with the middlebrow in terms of taste or aesthetic judgment; they just ignore it. Yet inherent in that ignoring is a clear aesthetic judgment; these cultural forms indeed constitute a vast wasteland, and people who consume them are probably not that capable of refined aesthetic judgment. Could it really be that apparently politically-aware cultural studies scholars agree with MacDonald in his condemnation of "tepid, flaccid Middlebrow Culture?" Woe to the cultural critic who admits to being a "scholar-fan" of *Touched by an Angel*!

Second, cultural studies scholars may also decide to analyze popular media in terms of their success or otherwise in representing reality, or in terms of how they maintain or challenge dominant ideology—an analysis of representation of African-Americans on *ER*, a discussion of gender roles on current sitcoms, the subversive potential of *The Simpsons*, and so on. A show like *Ellen* becomes significant not in terms of whether it is well done, but in terms of its ground-breaking role in depicting gay people. Once again, the question of "quality" is ignored, and the assumptions about viewers (who may or may not be directly addressed) tend to focus on whether they are being taken in or fighting against the prevailing hegemony. It is an essentially sociological approach, exemplified by critics like Newcomb (1974) whose perspective (as paraphrased by Kellner 1999) is that television narratives "orient people toward contemporary social reality,... articulate and help resolve conflicts, and... provide a steady stream of commentary on existing society" (p. 62).

The third major theme in cultural studies is of course the empirical study of media audiences, specifically through ethnographic approaches like those I have taken in this book. The most anthropologically-inspired approach of all, it also avoids aesthetic considerations. Although scholars rarely expound on their own relationship to the genre under study, this approach seems to be used mostly when the scholar is *not* a fan, with the notable exception of some feminist studies, which often self-consciously seek to reevaluate "feminine" genres such as soap operas (Brown 1994). At issue is not quality, or even ideological representation, but rather how an audience "reads" the text socially and ideologically. "Although frequently informed by a desire to investigate, rather than judge, other people's pleasures, this very avoidance of judgment seems somehow to recreate the old patterns of aesthetic domination and subordination, and to pathologize the audience" (Brunsdon 1990, 69).

As an anthropologist by training, I have never felt comfortable taking the scholar-as-fan approach; in fact in my one foray into a personal enthusiasm (Bird 1994), I made a conscious decision not to address my fan status, and have since wondered if my approach was fundamentally dishonest. Rather, I have felt generally quite comfortable with the second two approaches. I am interested in cultural processes, how they operate, and how they are changing, and I am especially interested in the media's role in the construction of cultural meaning. In the various studies discussed in this book and elsewhere, my goal has been to understand how media are integrated into everyday culture, and how people interact with and through media. And all the time I have been doing this, whether my focus has been on tabloids or television, I have been nagged by that annoying question: "But is it good?"

Ethnographic studies (my own included, I hope) have made it abundantly clear that viewers and readers are much more active than the old doomsday critics of mass culture could have ever believed, even within undeniable ideological and economic constraints. However, as Frith (1991) points out, we actually know very little about how audiences make *aesthetic* judgments, even though "what we're talking about is pleasure ... [which is] first and foremost connected with the *fictional* nature of the position and solutions which the tragic structure of feeling constructs, not with their ideological content" (Ang 1985, 135, italics in original).

Perhaps in our arrogance we assume that issues of taste and distinction are as irrelevant to audiences as we have made them for ourselves, when studying genres we do not especially like. I became acutely aware of this when I stumbled upon the decidedly middlebrow *Dr. Quinn*. My interest in the show was focused almost entirely on its representation of American Indians, as part of a larger study of such representation across popular media. I analyzed this representation, and I was also delighted to encounter some American Indian readings that definitely qualified as "resistant" (Bird 1996). But it was only when I became involved with the electronic discussion list that I began to appreciate that a text I had essentially ignored (apart from the Indian dimension) was the focus of rich and absorbing *aesthetic* analysis that was different in quality from the more sociological or identity-building discussion that also took place (see chapter 3), and which is more usually the focus of audience studies. At this point I began to appreciate the paucity of studies of such processes; in fact Frith points to only one significant set of data that explores, as he puts it, "the popular aesthetic in action" (1991, 112). This is the information collected by the 1938 and 1943 Mass Observation studies of British film tastes. These studies found that audiences have clear criteria by which the quality of popular films is assessed, including technique, believability in terms of the expectations of the genre, and creation of empathy and emotional response, a conclusion confirmed in McKee's relatively rare study of the aesthetic judgments made by fans of the British "cult" television show *Dr. Who* (McKee 2001).

Invoking Bourdieu (1984), Frith talks about different discourses that seem to operate when discussing the ways in which people make cultural judgments. The sharpest contrast is between an art discourse and a pop discourse, although he also considers a "folk discourse." In the first, a piece is judged against an ideal of transcendence – that it rises above the everyday, transcending the body and specificities of time and place, while the pop discourse values fun, routinized pleasures and desires, and legitimized emotional gratification. As Bourdieu claims, high art rituals are an aspect of bourgeois "distinction," whereby high cultural experience is defined as purest. However, as Frith points out, Bourdieu apparently accepts the idea that these different discourses actually do separate people rather neatly by class, and that we can make assumptions that, say, popular TV viewers are consistently operating within a folk or pop discourse. In Bourdieu's view popular taste is "the exact opposite of the Kantian aesthetic" (p. 5), subordinating form to function and closing the door on the "popular" ability to assess a text other than in terms of what it "does" for the consumer, which is often seen in terms of emotional affect. This is an assumption shared beyond the circle of academic critics; for instance Sparks (2000) points to the taken-for-granted view in Britain that "tabloid readers" are qualitatively entirely different people from those who read "quality" papers. Frith argues that Bourdieu's critique allows cultural studies analysts to accept uncritically that "the people's pleasure" is all at the level of the "pop discourse," ultimately a patronizing position to take, and one that draws a very clear line between "them" (the objects of our study) and "us" (the cultural critics). In fact, of course, the art/folk/pop boundaries are inherently unstable and the aesthetic judgments we all make draw on all discourses—the body and the spirit, pleasure and moral interrogation, sense of place and timeless truths.

Where does this leave the question of aesthetic value and cultural studies? Like McKee (2001), I agree that most scholars are not actually denying the possibility of making some kind of value judgments. Many mention that the audiences they study do just that, yet the actual process remains a mystery. Hobson (1982) in her classic but consciously atheoretical study of British soap opera fans, states bluntly that "the viewers are the critics. Or at least the only ones who should count" (136), but her characterization of these viewers does not offer a picture of them as "critics." Instead, she focuses on how they use the program in their daily lives. This is a pattern we see repeatedly, as scholars seem unable to take that next step, perhaps because the basic sense of empathy is missing. We know that these "fans" think highly of their favorite texts, but we cannot really empathize because we cannot see the texts through their eyes; perhaps audience studies sometimes do "become the branch of sociology that Robert Warshow was afraid of: an investigation of more or less revealing responses on the part of an audience whose responses and interests and yearnings we are fated to study and never to share" (Gould 1999, 134).

As I became more involved in the DQMW list, I was indeed fascinated by the members' diverse interpretations and wide-ranging discussions. Yet at the same time, I gradually became aware that I was starting to evaluate the program itself differently, seeing it as distinctly "better" than I had previously thought. Frankly, I never truly became a fan—I make no effort to follow the series now that it is in syndication—but I did start to enjoy it more than I ever expected. I found myself defending it at academic conferences, when colleagues made good-natured jibes about my journey through schlock, from tabloids to *Dr. Quinn.* I began to think more deeply about the rather obvious point that aesthetic value, far from being intrinsic, is learned. As Higgins and Rudinow (1999) point out in relation to music, it is necessary to acquire some cultural capital when approaching any genre, whether it is classical or rock: "To appreciate music of either kind, one must be or become a participant in the relevant musical culture" (p. 116). Davies (1999) makes a similar point, arguing that "an aesthetic interest does not vary from genre to genre. Many aesthetically important properties—such as narrational, representational, and expressive ones, or others such as unity in diversity—are common to many genres, periods, or styles . . . " (p. 202). Gracyk (1999) agrees, and makes the significant point that an understanding of the aesthetic evaluation process for any genre does *not* mean that we then have to make a judgment about whether that genre is truly "art." I believe it is the queasiness about making "artistic" judgments that is partly responsible for the sidestepping by cultural studies scholars when it comes to questions of taste and aesthetic appraisal.

This realization opened the door for me to show how ethnographic reception study might be helpful in understanding and even making judgments about *quality* in popular media, without falling into elitist distinctions. I do not have to say that *Dr. Quinn* is "art" to suggest that it embodies qualities that allow it to be a richer text than, say, *Beverly Hills 90210.* So at this point, I would like to again enter the world of the *Dr. Quinn fans,* but now with a focus on their role as appreciative, discriminating viewers. In addition, I will amplify this discussion by adding the voices of daytime soap opera fans, with a goal of showing how aesthetic distinctions are both similar and different among these two fan communities.

DQ MEETS *OLTL*: AESTHETICS THROUGH THE EYES OF THE VIEWER

Chapter 3 introduced the e-mail discussion list, DQMW-L, which I happened upon by chance. When first invited to listen to the list's discussion of my academic writing, I was skeptical; I suppose that like most of the critics of the show, I expected the "fans" to be "typical TV viewers" who were more interested in Joe Lando's chest than a consideration of character development or moral issues. I was wrong.

The second group, which I do not know nearly as well, is a message board (www.mediadomain.com/oltl) for the daytime soap opera *One Life to Live*, a program that has survived more than thirty years on ABC, and which has fans who can remember storylines from its very earliest days. Coincidentally, it was also the program on which Joe Lando got his major career break, as fan favorite Jake in the early 1990s. Soap operas and their fans have been studied quite intensively; most recently by Hayward (1997) and Baym (2000), who studied the formation of community on an online news group. Both pay some attention to the judgments made by fans, although their main focus is once again functional.

In the interests of self-disclosure, I should point out that I have been a somewhat dilatory viewer of OLTL for many years, going through periods of interest followed by years of neglect, but have found it quite easy to become updated on the long-drawn-out story lines. Recently, I began monitoring the board again during a rather interesting period in which the fans were regularly expressing displeasure with the state of the show. The animated discussion of what was *wrong* with OLTL was useful in understanding what makes a "good" soap opera. Very clearly, there are some evaluative standards that are shared between fans of the prime time and daytime programs; it is equally clear that there are other criteria that are quite different.

Before beginning the discussion, I should also point out that the two forums, although both electronic, are different in character. The DQ e-mail list clearly functions as a community, whose connections go beyond the show itself, and members post openly and using their own names. The OLTL message board has only quite rudimentary features of a community, members all use pseudonyms, and discussion is limited to the show itself. Recently, a new member posted a message asking others to write a little about themselves; almost 50 did so, and many expressed pleasure at this different idea, but it has not led to a more personal level of discussion. Postings can be quite long, reaching hundreds of words occasionally, but they have never reached the prodigious lengths of many DQ posts.

THE PLEASURE OF THE TEXT

In both forums, there is plenty of talk about pure pleasure and desire. DQ fans admire the male leads, and discuss them endlessly. At one point, news of Joe Lando's haircut (damaging his image as the disheveled, long-haired mountain man) swept through the list like wildfire, and fans debated the pros and cons of his new look. But there is also a self-consciousness about this aspect—posts on such pressing topics as Lando's hair are often flagged—"Warning: Shallow Post Follows!" Members actively enjoy parodying themselves as besotted fans. Similarly, the daytime fans (who include many active male posters) discuss the appeal of the OLTL actors, although there is also a very vocal thread that critiques producers for assuming that pretty former models of both sexes are enough to maintain fan interest. There was endless lampooning of recent storylines that had one well-built male actor constantly peeling off his shirt in every scene (he spills coffee, he is working out when someone arrives at the door, and so on). Fans point out that, because most tape the program and can fast-forward dull scenes, intriguing storylines and character development are essential to hold their attention. Nevertheless, the physical appeal of the actors is a constant source of discussion, with regular polls about who is the "sexiest," "most beautiful," and so on.

The DQ fans, however, go beyond their admitted physical attraction to the stars. Amid the joking and lusting, we find careful analysis of their own desire, for instance in a discussion of how Lando's character Sully's hair is symbolically important because of the way it ties into the mythical history of the type of character he is portraying. As one lister writes:

> Sully is a mix of a Hawthorne-like hero and female fantasy for us . . . In traditional western boy fantasy the hero expresses himself by shooting bad guys. In female-led Quinnland our hero expresses himself with . . . vision quests and feelings . . . When Sully is dismissed by male smart alecky critics as Fabio on the Prairie they only show they don't get it . . . the Hair Matters Of Course but the most important part is . . . the CHARACTER of Sully.

This was among a flood of long, detailed postings that dominated the list in 1999, after the program's cancellation. With episodes now only in re-runs, the list became more focused than ever on critical but pleasurable analyses both of the television texts and the fan-fiction texts regularly posted. In moments like these, we can clearly see the DQ fans in the role of "fan-as-scholar" discussed by Hills (2002): "It remains a significant and under-researched fact that fan-scholars have directly drawn on academic knowledge in order to express their love for the text" (p. 19). This particular critical thread addressed the question of "the hero," and focused primarily on the characters of Sully and Hank, the tough-guy bartender and brothel-keeper. Listers very deliberately contextualized the characters not only within the genre of television drama, but also within literary traditions, such as in the Hawthorne reference. One poster introduced the notion of the flawed Byronic hero, and there is debate about whether this idea applies better to the aggressive Hank than to pacifist Sully:

> ...the [Byronic] character is very complex—and very specific. He has a "noble" background and family connections, but usually is found as a leader among the less savory elements of society. He has left his home, is an outcast in many ways, and is often misunderstood or mis-read by the "good" characters. He has a past—a secret which drives him. It is more than sad; in some way it is usually scandalous. And this hero is saved or redeemed by his own code of honor (which never quite matches up with society's) and by his ability to love. But regardless of what else happens, he still considers himself "damned." And the tragic nature of the character comes from his excess, as seen in Heathcliff (in *Wuthering Heights*) and in many modern manifestations linked to the Byronic.

Is Sully a "pure hero" or a "Byronic hero," as might be implied by his first name (Byron)? Yet another literary reference surfaces:

> An old distinction about drama colors my view of heroes...In a tragedy...all threads wrap around the hero...his fate is the fate of a nation, so one overlooks just how self-concerned that makes him. In a comedy, in the broadest sense, one still has central characters, but they are instead community members, and the best of them are sort of noble but in a manner that expends their unique identity on the well-being of others— they "flow outward" rather than pulling others inward...Both sorts of characters appear in Melville's fiction, especially in Ahab and Ishmael, who in every way display these centripetal (in Ahab's case) or centrifugal (in Ishmael's case) tendencies. Ahab...is certainly Byronic—Ahab draws his crew into an impossible chase even as he is riddled by the belief that he is damned, readily throwing a NO to the heavens. Ishmael follows all belief almost as a reader rather than a participant...Does this distinction work when viewing Hank and Sully? Both have had deeply Byronic traits...and in retrospect have had long periods of self-doubt, even self-damnation. But Sully is much

more the Ishmael, who browses at the tables of many beliefs . . . he must remain enigmatic and outside of the community in some sense always . . . Hank on the other hand has, amazingly enough, given his all for this literal community of Colorado Springs, despite the insight into its hypocrisy . . . At any rate, I feel certain that the two of them are crucial poles in the multiple web of ideas about heroes that Beth Sullivan designed for us . . .

Or perhaps Sully is a "hero" in the mold of a cinematic John Wayne? No, most agree:

> John Wayne's movie characters are considered almost the paradigm of heroic and yet are very different from our DQMW heroes, particularly . . . when it comes to sensitivity. I think that . . . is what makes Sully resonate so strongly with the female population . . . Sully takes the hero one step further to the category of ROMANTIC hero. Sully particularly puts the woman he loves before himself and that is pretty unusual for a male . . . To me John Wayne's characters must be the ultimate GUY-type heroes, lots more guns and guts than a Sully type. But there are similarities too, particularly in that critical "code of honor."

Other posters weigh in with considerations of the Western hero as defined by Cawelti (1976) or Tompkins (1992), or the archetypal hero of Campbell (1949). Members analyze specific scenes, sometimes transcribing long script excerpts to make their points. One such example was a scene in which Sully and others have chased down a man (McBride) who has killed his wife, and now goads Sully into killing him, while Robert E., the central African-American character, attempts to calm him:

> This scene, with the close juxtaposition of Sully's sense of lost order, Robert E. insisting Sully maintain his rules, and then the powerful symbolism attached to the drum is brilliantly done. And the drum itself is a very complex symbol. Not only does the sound we hear literally mimic that of the heartbeat of Sully's unborn child which is also threatened at this point, and which is perhaps, his most compelling emotional, spiritual, and even biological connection to order and continuation, and a physical expression/result of faith in that order, but also the drum in any music helps establish and maintain order . . . And here it is being misused by McBride (one who has clearly lost all sense of order) as a means to disrupt the rescue party's common sense and taunt them into breaking discipline and coming after him . . . Sully restores that order by refusing to respond as McBride wishes. But

his ability to maintain this order is based on the fact that he is also able to understand and interpret McBride's crisis and loss of faith. Sully is powerful because he accesses both.

It is in discussions like this that we see how the fans develop an aesthetic that is intrinsic to the text of DQ—that their heroes may be understood in light of many other generic conventions, but that there is also a logic and structure to the program itself that knowledgeable enthusiasts follow. It is that recognition of the "correctness" of a particular aesthetic that also allows fans to critique each other's fan fictions. The best stories reflect a deep knowledge of the show's conventions and history, while adding new layers, often to relatively minor characters. Thus one member comments on another's story about Hank:

> It was like watching a photo develop. A black and white photo that you can watch coming up on the page. The faint gray traces and outlines that I was so familiar with filled in with such texture and richness that I was left with that feeling of "magic" that still strikes me as a dramatic image gradually appears from misty nothing-ness, and then drifts gently in the tray, glowing in the darkness, and waiting to be moved to the "fixer" so the sharp reality of it won't fade. I was stunned. There was no mistaking that what I had read WAS Hank in some fundamental way. Completed in a way that we never saw, probably never would have seen in Colorado Springs, if only because the "light" of the main characters made it easy to overlook the rest to some degree.

The very "literary" style of the DQ fans is not regularly matched by the OLTL enthusiasts, although it does happen. OLTL posters are much more self-conscious about over-intellectualizing and often flag such analysis with somewhat defensive comments. For instance, speculation greeted a new character played by a familiar actor, and thus assumed by long-established viewers to be evil cult leader Mitch Lawrence, who had sup-posedly died in the 1980s. He was now calling himself Michael Lazarus, and some were wondering if he was not Lawrence but an identical twin (given that the "twin" ploy is just as much a soap cliché as the "return from the dead" story). One fan disagreed, pointing to the clue given in his new name: "I think it is meant to have a certain irony. Michael was the Warrior Archangel—the Avenging Angel. Mitch said today (and in the flashback) 'Vengeance is mine, sayeth the Lord.' So Michael Lazarus is an ironic perversion of Michael, God's avenging angel, and Lazarus, Jesus' friend who he raised from the dead." The poster immediately fol-lows the comment with a self-deprecating remark about "showing off," in the same way another poster ends a long comment with "Forgive the spelling, I'm trying to set the record for pretentious references." And even as they spend a great deal of time satirizing the show, OLTL fans do take their genre seriously; in fact their frequent parodies usually function as quite biting commentary about what happens when soap opera is done badly.

BEING TRUE TO THE GENRE

Indeed, one of the key elements in aesthetic appraisal of narrative is a consciousness of the "appropriateness of narrative" developments. Like any traditional English literature analysis, DQ list members spend a great deal of time discussing character development on the show, against a background of knowledge of its history and conventions. For example they have discussed Jake, the town barber/surgeon, who begins as a weak-willed bigot, but evolves into something else:

> I think "redemption" for Jake begins to happen earlier in the season during "The Incident" episode. Jake responds at one point to Sully, "One Indian. I didn't know there'd be such a fuss. Indians die all the time. No one gives a damn." That episode is about the nature of empathy. Jake sees . . . and is frightened by the grief, anger and hatred that surrounds him . . . a seed becomes rooted and it changes Jake.

Indeed, a large amount of discussion focuses on how characters develop, within the context both of individual episodes and within the larger arc of the developing "story" of the entire series. As one member commented, "DQ attempted to have a psychological progression within it, while managing to have each episode concluded in itself. At the end, the characters have evolved in a mostly meaningful way, and we can still recognize in our mind independent episodes." There is approval when characters show growth, and disapproval when characters act inexplicably. By 2002, members were discussing episodes very much within a broad framework:

> . . . with the perspective of the entire series behind us, I can appreciate the episode on its own merits . . . Michaela showed growth toward Matthew [her adopted son] . . . She is clear on where she stands with regard to gambling, but she doesn't forbid him from doing it . . . She reassures him of her love . . . It was nice to see the Reverend get some background and more depth to his storyline. I wish we could have gotten more of that background as the series

progressed . . . The eagle storyline subtly tied into the main storyline . . . Only when Matthew had found his way was the eagle able
to fly away and find his own way. Nice parallels there . . . Finally,
there was a continuity with past episodes . . . and foreshadowing
of future episodes when Robert E. mentions that he and Grace
are saving up to buy a house.

Members' strongest ire is reserved for "lazy writing" or poor character
development, exemplified most thoroughly in the passionate denuncia-
tion of the character Daniel, introduced late in the series and widely seen
as an attempt to replace Sully. Some of this clearly stemmed from the
choice of the actor John Schneider, whose main claim to fame was a lead
role in the defunct series *Dukes of Hazzard,* which was largely dismissed
as a "trash" show by DQ listers (again pointing to the aesthetic "sorting"
that belies the lumping of all series TV fans in one category). But most of
all, the listers agreed that the character was superfluous, had no organic
roots in the story, and had been thrust upon the reluctant fans:

> Daniel just solved problems alone while Michaela was depicted as
> looking on him in some sort of semi adoration. Sully brought out
> the best in Mike in terms of her passion and spirit and desire to
> help others. Daniel seemed to reduce her to the stereotypical help-
> less female and he would take care of hers and apparently everyone
> else's problems as well. Daniel reminds me of the android or robot
> hero in fictional stories who seems perfect because he has no faults
> but ultimately finds that his lack of faults is the very thing that
> keeps him from being human. He seems to be there for other
> characters to react to either negatively or positively but doesn't
> seem to really have any real feelings . . . of his own. Ultimately, you
> couldn't hate Daniel since there was nothing of any real substance
> to hate.

Brooks (1994) writes that aesthetic awareness comes from a sense of
"not only *what* a text means but also *how* it means, what its grounds
as a meaning-making sign-system are" (p. 517). *Dr. Quinn* fans have a
keen sense of what the sign system of the show is all about, expressed
in exasperation when the coherence of that system is violated. Many believed that in the show's final two seasons, the producers were trying to
appeal to a younger audience, and that quality was sacrificed: "If a character you perceived to act one way, suddenly—with no understandable
motivation—begins behaving in an entirely foreign way . . . It's as if the
foundation has suddenly dropped out of the story." Similarly, when the
show was syndicated on "family-friendly" Pax TV and then Hallmark, it

was heavily edited to remove reference to the prostitution in the town saloon, with every use of the word "whore" deleted, giving the impression
that Hank was merely running a saloon, rather than a brothel. Pax also cut
out some harrowing scenes of death and suffering in an episode focused
on the historical massacre of Cheyenne at the "battle" of Washita—cuts
that according to one lister "totally destroy the power and poignancy of
the story . . . the cuts are almost a blasphemy." Members frequently discussed how such cuts were made "on ideological rather than narrative or
artistic grounds," which "destroys the underlying tension of the whole
storyline."

Meanwhile, what about the soap opera? DQ fans' point of reference
for "quality" seems to be the conventions of the show itself, seeing it as
its own distinctive body of work. OLTL fans reference "appropriateness"
by how things fit into the specific genre of the soap opera. They (at least
those on the mediadomain board), do not tend to be as self-consciously
reflective about their genre, in the sense of writing what amounts to
critical essays—although this does happen, as I shall show. Part of this is
inherent in the nature of the genre. Soaps provide 45 minutes of story five
days a week, every week of the year, and there is a sense of continuous narrative flow that makes it less easy to discuss particular moments. However, discussion reveals that there are very distinct conventions about
what makes a well-constructed soap, and fans discuss these regularly.
Furthermore, critical analysis has become much more feasible with the
time-shifting capabilities of videotape, and the recent development of
"Soapnet," an all-soaps cable channel that reruns great moments in soap
history. Extensive Websites also feature archives of stories, both on video
and in print, and fans create places where particular storylines are iso-
lated and discussed. In other words, technology has allowed the soap
genre, as well as the episodic drama genre, to escape the ephemeral fate
that long hampered fans' ability to develop a more mature aesthetic
appreciation.

Occasionally, the mediadomain fans do embark on a debate about
"quality," and show they are knowledgeable and thoughtful enthusiasts.
In September 2002, a member posted a query asking people to define
a "good soap," following a large number of posts about how bad the
show had become. There was general agreement that "character" was the
key.

> Character-driven stories . . . Plots come and go, and they soon be-
> come repetitive because there aren't really all that many of them,
> but characters can organically develop indefinitely. I keep coming
> back to that three and a half hours of airtime a week to fill. You

> just can't keep surprising or shocking or amazing an audience all that time... But nobody minds seeing an old friend for 3 $\frac{1}{2}$ hours a week.

Indeed one of the central themes in the hatred directed at the producers during this period was the reliance on "plot-driven" stories, where a character would suddenly be introduced out of nowhere. For instance a doctor suddenly acquired a never-before-mentioned younger sister, who apparently appeared for the sole purpose of bringing up the "secret" about the man's dead wife in scene after scene with him. With no history, motivation, or story of her own, she was regarded as tedious and pointless. As one poster comments, "No story, no plot is interesting in itself. A story or plot is only interesting as it reveals the possibilities of character, by showing characters in action." Cohen (1999) points out one of the motivations behind critics' derision of popular culture: "They suppose that it is made formulaically, calculated to appeal to a certain group, and that it is invested with no personal conviction by the artist, if indeed such a manipulator deserves to be called an *artist*" (p. 140, italics in original). This assumption also implies that viewers do not have the discriminatory abilities to perceive this manipulation. Soap opera fans are very aware of the formulaic nature of their show, yet it is clear that they are indeed angry when they feel they are being taken for granted or insulted by poor execution of that formula.

Members regularly discuss the validity or consistency of characters, often with some passion, as in a favorite target, the character of lawyer Sam Rappaport. Without going into detail about his history, I need only say that he has been problematic from the start, first because he was given enormous amounts of airtime immediately, and a cardinal rule is that characters should be introduced gradually. Later he evolved into a rather bland "do-gooder," who seems to be peripherally involved in many storylines. In October 2002, a regular member posted "the case for Sam," in which he asked why this character, painted as an all-around good guy, was so disliked. The torrent of replies echoed the condemnation of DQ's Daniel:

> You have wonderfully depicted Sam as a good person, a true Prince Myshkin [from Dostoevsky's *The Idiot*]. Yes... "good" characters can be made interesting, but only if the writers take care to make them "active"—characters who do things rather than having things done unto them. Sam isn't active. He can only be said to do anything in the sense that a useful inanimate object may be said to do something. Penicillin cures infections, Sam clears up the dead baby lie. But it's not very interesting to watch either of them do so...

These discussions show that there are clear similarities between DQ and OLTL fans' criteria for quality, and I would suggest that a concern for coherent character development is central to appreciation of genres from TV to literature. At the same time, there is a clear awareness that the genres *are* different. DQ fans, like many fans of prime time dramas, have a clear sense of "realism." They know that DQ is not "real" and they also happily accept anachronism, such as the focus on 1990s values that often seems out of synch with the 1870s. But they expect basic accuracy; they dislike it when Michaela performs surgery that was not widely known at the time, and will have long (if often tongue-in-cheek) debates about such details as a "mare" that is quite obviously male. They enjoy the incorporation of real people and real events into storylines. Soaps require a different order of suspension of disbelief. Colorado Springs in 1872 has a certain sense of concrete reality, linking it to the real world, even though fans are perfectly aware that the town was not literally like this. Llanview, Pennsylvania, the home of OLTL, lives in a kind of suspended reality. Known to be a small town, it nevertheless has featured a major university, two large, competing newspapers, extensive medical facilities, national television studios, and so on. Characters may change their profession from, say, television producer to police chief with minimal explanation; small children disappear for a year and reappear as teenagers (known by afficianados as SORAS, or Soap Opera Sudden Aging Syndrome). Yet viewers understand all this and even embrace it, as long as there is a recognizable internal consistency:

> The heart of any soap is a group of characters: clearly defined characters, with lives and concerns that in some way touch us, and with whom we can identify—by love, hate, admiration, pity, whatever. These characters must inhabit an identifiable place: a street, a village, a city, a hospital, even a haunted castle—the ghouls and vampires of *Dark Shadows* led lives that were in their own way as mundane, as identifiable, as the characters of *East Enders*. They have to belong there ... have a credible function ... Viki is a successful character: a newspaper editor whose interests and sympathies reach deep into the community of Llanview.

Viewers understand that soaps are not "real":

> Soap is a genre in itself, and it is essentially melodramatic. That's not a criticism or slight—some of the greatest movies and novels are mellers. It just means that while lives may be portrayed to a degree realistically, there is an emphasis on the emotional and interior lives of the characters, and these tend to be drawn vividly ... the stories often pursue possibilities way beyond

anything that might happen to those of us who inhabit the real world—but they must not exceed what believably might happen to these characters *within their particular world.* In the soap world, they do things differently, they don't have our constraints.

Another fan also compares soaps with other genres:

> I can't help think a little about Grand Opera...of highly dramatic emotional moments often at their peak when hit at High C. Soap operas IMO also depend greatly on those highly dramatic moments that are best when given a slow build-up leading to a crescendo that leaves the audience spent emotionally, but with the feeling that they've experienced something really special. As an example...the final scene in *Madam Butterfly* in which the Japanese madam discovers that her American soldier is not returning to marry her and take her and their love child back to America, but that he is already married and has just returned for his son. Devastated, she dresses the little boy in a sailor suit and sits him on the steps to wait for his father and puts a little American flag to wave in his hands, while she quietly goes inside the house and kills herself, singing her heart out at the end while the little boy is gleefully waving his American flag... My favorite example of high drama on OLTL was Megan's death, especially the tense moments when Jake and Andrew were bravely facing all kinds of danger to get him to her side in time to say goodbye...

THE MORAL AND EMOTIONAL AESTHETIC

When we explore the conventions of both daytime and prime time drama, we see that fans do apply a kind of formalist aesthetic principle, if indeed "art is the imposing of a pattern on experience, and our aesthetic enjoyment is recognition of the pattern," as Whitehead (1943) said in a widely-quoted sentence that has become almost a cliché. The self-conscious, explicitly critical and comparative approach of the "fan-as-scholar" is a much more normalized feature of the DQ list than of the OLTL board, but it does come into play there when stimulated. For DQ fans, the standard of imitationalism is also a factor—the theory that the success of a text is also traceable to how well it reminds the audience of the real world—while this criterion is far less crucial for soap fans.

However, in both groups, perhaps the most dominant form of aesthetic appraisal is that known as emotionalism—a theory that emphasizes the expressive qualities of the object. How well does it communicate moods, feelings, and ideas? Some philosophers find this approach problematic, touching on such dangerous areas as sentimentality, of which the standard view is that it is "a character defect, not just an aesthetic flaw" (Knight 1999, 414). Certainly both DQ and soap operas have been decried for their "easy" sentimentality. Yet both sets of fans show that in this area too they are discriminating and active in the way they respond, usually connecting the emotional and expressive with the moral dimension; a "good," effective emotional response is one that connects with an idea about moral appropriateness. In attempting to define the aesthetic in a way relevant to cultural studies, Hunter (1992) emphasizes the role of art in exploring moral issues: "It is not that literature is open-ended, but that we open its ends, subjecting it to permanent aesthetic surgery as a means of operating on ourselves" (p. 365). DQ fans clearly do this all the time, with their extended discussions of moral issues raised in the show (see chapter 3). They make clear judgments about the quality of different storylines in terms of their ability to evoke emotion through moral debate. "Light" episodes are appreciated, as in a 2002 discussion of a rather unlikely episode in which a circus comes to town, and various regulars take part. One poster wrote: "It's a funny episode and worthy for a lot of things: the interaction of the characters, the lovely metaphor of Mike and Sully's jump, the theme of diversity," and others agreed. But it will never be seen as a "great" episode, partly because of its implausibility and partly because it lacks "layers" of moral complexity and emotional depth.

Much different was the discussion of the Walt Whitman episode. At one level, there was the debate about the morality of homosexuality, which led into a more complex discussion of moral issues and tolerance. At another, listers explored characters' responses in terms of expectations created by past experience. At yet another level, members discussed Whitman's poetry, how it was incorporated into the story, and how it evoked a strong emotional response. As a member writes:

> I have remarked in the past of the associations I have with the working-class imagery in Whitman's work and the character of Sully . . . Here, again, we have that type of visual reference from *I Hear America Singing*: "The carpenter singing as he measures his plank." It is noteworthy that the visual pairing only occurs with Sully . . . Sully may not have read Whitman for us, but he typified the spirit of the poetry's images.

Another adds: "Whitman . . . was the first 'modern' poet and as such took the risk to defy the conventional with his use of words and form. His vision still allows us to experience life in all its richness and beauty and to glorify the mundane as part of our own fabric of living." Somehow it seemed rather fitting that Whitman, these days definitely "high" culture, was being exalted in the "low" culture setting of the pop TV show. After all, he wrote in *Democratic Vistas* that culture should not be "restricted by conditions ineligible for the masses," (quoted in Twitchell 1992, 65), a point not lost on one lister who remarked, "Perhaps Whitman's Jesus-like figure was and is the return of the repressed, in every sense." Through-out the Whitman thread, and in fact throughout most of the list debate, there is a widely-accepted view that the "best" episodes are those that con-tribute to the characters' moral progression and growth, and which have multiple moral layers. For instance, most episodes have a main story and a "B" story, and those that "work best" are the ones that inter-relate to inter-rogate similar ethical themes. And in turn, the best morality tales are also those that evoke strong emotions, a point echoed by the soap opera fans.

The OLTL fans also place high value on emotional response to situations that deal with issues of morality. The moral code of soaps is certainly different from that of DQ; casual sex is not rare, while lying and deception are often keys to moving storylines along. Yet at the same time, viewers have underlying expectations of how a good soap handles morality, and the emotions that go with it. The most prevalent and vehement complaint about the 2002 incarnation of OLTL is that it was full of couples falling into "insta-love"; their growing attraction was not shown believably, and they fell into bed barely knowing each other's names. Fans loathed storylines that showed, for instance, a male "hero" lead having a one-night-stand with his fiancée's mother while the couple were temporarily estranged, resulting in the mother's pregnancy. The disaster was compounded when the resulting embryo was transplanted into the daughter, who became a surrogate for her own mother's child, fathered by her boyfriend. Truly a bizarre situation! But it was not so much the bizarreness that enraged fans, who will take medical impos-sibilities and even paranormal events in their stride. They were angry because they found the story repugnant and degrading, deriving from out-of-character behavior and manipulative, plot-driven writing. They considered it along with a series of other "nasty" storylines; as one thread asked, "Could this show BE anymore insulting to women these days?" Copious discussion addressed stories that depict OLTL's women as "love starved," "obsessed," "abuse-sucking," "meddling hag," and "sado-masochistic." As one writes, "Don't [the producers] understand their core audience?" Their basic argument is that such stories lack any sense of moral or emotional realism, and instead of evoking empathy, they evoke disgust.

Beach (1997) dismisses the idea of a moral aesthetic as too vague, asking, "How does one situate oneself critically vis à vis a work of art in such a way as to perceive an ethical or social function of the work that is separate from its status as text or object?" (p. 98). It seems to me that this is indeed an important question in some contexts—do we judge *Birth of a Nation* as a groundbreaking and brilliantly-executed film, or as a repugnant, racist diatribe? But for most people, most of the time, whether contemplating a great novel or a TV show, the formal, the moral and the emotional are not separated, but (if effective) work together to create the sense of pleasure and aesthetic appreciation. McKee (2001) quotes a fan writing about a dramatic, climactic moment in the sci-fi series *Dr. Who*: "If the closing soliloquy, as the Doctor and Ace walk off to further adventures, doesn't bring a lump to your throat, then there is no poetry in your soul." DQ and OLTL fans hope for that same moment of poetry when the ethical and emotional come together with meaning that speaks to them directly.

Finally of course, media fans also value the quality of basic features like acting performance, directing ability, sets, writing, and so on. McKee describes how among *Dr. Who* fans, "Particular formal elements of the text (performances, writing of dialogue, construction of plot, framing of shots, editing pace) are evaluated, and assigned a value between 'brilliant' and 'awful,'" and DQ/OLTL fans do the same. Standards differ; soap fans understand the constraints of limited sets and daily taping, and especially value achievements that transcend those constraints. Acting is also evaluated differently; for instance, soap fans happily acknowledge that "good" soap acting may be very different from that in prime time TV or films. It tends to be more melodramatic and "over-the-top"; although overindulgence is often dismissed as "chewing the scenery," fans also reserve a special, campy enjoyment for melodramatic, "crisis" scenes in which major secrets are revealed and actors pull out all the stops in a style that seems to work only on soaps. Recently, fans discussed the fact that many small roles are played by New York stage actors (where the show is taped), opining that the more "stagey" style of acting fits the soap genre well. DQ fans prefer a more understated style, although they also enjoy "big" moments, such as when Michaela delivers a speech destroying an evil opponent. But the fans are united in their disdain for wooden, expressionless acting, and especially for "models-turned-actors," clearly chosen for their looks alone. While DQ fans maintain restraint in critiquing bad acting, they do acknowledge it; OLTL fans are ruthless in their lampooning of acting ineptitude.

CONCLUSION: WHITHER AESTHETICS?

McKee (2001) states, "I am aware of little Cultural Studies work on how value judgements are made outside the academy, in the everyday world."

This chapter has been an attempt to explore this issue, and to suggest that this process is more complex than many understand. My discussion directs attention away from the text as compared "objectively" with any number of other texts, instead focusing on the active process of meaning-making engaged in by knowledgeable audiences. As Willis (1998) suggests, this is a "grounded aesthetics" that "places *social practice* over the veneration of *things*" (p. 175, emphasis in original).

Am I myself a fan of *Dr. Quinn* or *One Life to Live?* I think probably not; most of the time, I can take them or leave them—although I will confess to periods in my life when I was hooked on the soap and its characters. But thinking about the evaluative work done by fans leaves me with a deeper appreciation not only for them, but also for their chosen genres. Is *Dr. Quinn, Medicine Woman* art? Is it even good television? Can television be considered art, anyway? I do not think that is an especially relevant question, although it seems to be the one that people find most troubling when discussing television and other popular media.

Instead, I think careful audience study can help us refine our thinking about aesthetics and culture. First, assumptions are too often made about how "the people" watch TV, and what sort of value they attach to it. Television viewing is not one, uniform type of activity. Many of us "veg out" a great deal of the time, but we may pay special attention to particular programs and may engage in more thoughtful value judgments about them. There will be a host of aesthetic, social, cultural and political reasons for those choices, yet cultural studies has focused almost entirely on the socio-cultural dimensions. And as Frith writes, "To gloss over the ceaseless value judgments, the continuous exercise of taste, being made within popular audiences themselves is, in effect, to do their discriminating for them, to refuse to engage in those arguments which produce cultural values" (p. 105).

Furthermore, even critics who are trying to deconstruct the distinctions between "art" and "pop" styles of aesthetic valuations still fall into modes of thinking that force people into one category or another. Even McKee (2001), after detailing the many evaluative criteria used by *Dr. Who* fans, is puzzled by the fans' "retention of the language of aesthetic judgements," saying that "this use of aesthetic discourses . . . shows us that popular culture continues to use the discourses established for high art," even though they do not seem appropriate. He argues that this reflects the "bleeding of taxonomies developed for one kind of culture into the assessment of another." This kind of analysis paints a picture of popular culture fans as quite separate from "high culture" fans, and almost as inappropriately aping the high status language they have learned about "art." I think it would be more appropriate to see *Dr. Who* fans as much like DQ and OLTL fans; they enjoy a wide range of popular forms, and they tend to evaluate them in similar ways, but with tool kits and evaluative criteria that are adapted to each genre and dependent on a thorough understanding of each.

Critics who see the "art" and "pop" discourses as mutually exclusive often argue, with Bourdieu (1984) that art is appreciated with the mind and spirit, and pop culture with the body and the appetite. Audience study helps us clarify that aesthetic appreciation works at different levels at the same time, and it involves the mind *and* the body, whether the object is high art or popular media. As a teenager I devoured tragic novels like *Wuthering Heights* and *Jude the Obscure*, and I was transported by the power of the language and images. But I also "fell in love" with Heathcliff in a very emotional and direct way. Because DQ fans drool over Joe Lando does not mean that they are incapable of appreciation at a more abstract level, nor does it mean that a middlebrow show like DQ is by definition an artistic wasteland. As Cohen (1999) comments, "a great deal of the finest art we know appeals to various audiences in many different ways. The different constituencies within such an audience are not always divisible into high and low appreciators: sometimes the divisions are along quite different lines ... " (p. 141).

Furthermore, knee-jerk criticism of shows like *Dr. Quinn* and soap operas often betrays a common association of the feminine with vulgarity and excessive emotion, a point DQ listers understand and resent: "People who have never seen an episode will denigrate the show – 'What is this, some kind of Dr. Quinn-thing?' has specific connotations ... and is usually accompanied by a tone or a statement dismissing the importance ... of the emotional as 'girl-stuff.'" Indeed it is the very capacity of popular forms to elicit emotions that undergirds elite disdain, because they are believed to be "easy": The popular text "is intentionally designed to gravitate in its structural forms (e.g., its narrative forms, symbolism, intended affect, and even its content) toward those choices that promise accessibility with minimum effort, virtually on first contact, for the largest number of relatively untutored audiences" (Carroll 1997, 190). Undoubtedly, *DQMW* can be and is enjoyed by viewers who simply watch and give it little thought—but so can "great art." No matter what the genre, some audiences will be better educated about its subtleties and conventions, and will thus appreciate it more fully. As Shusterman (1999) writes, sentimentality and emotion are a trademark of such "low" genres as country music, and this emotion is dismissed as "cheap" and manipulative by critics. "Yet both scholars and singers of country music see its 'open display of emotion' as the key to its success and sense of authenticity" (p. 226). A DQ fan expressed her frustration at the characterization of Internet fans as "crazy," following reports that Jane Seymour and producer Beth Sullivan were wary of their enthusiasm:

> I prefer to think of this list... in the same vein as a book club (whose members) come together to discuss its implications, story-line, characterizations and analogies in a coherent and respectful manner... The fact that our story... is visual... makes us likely every now and then to wallow in the shallowness of a beautiful face, great body, good hair or other attribute...

She follows up:

> Methinks (they) have been reading too many Internet horror stories... if Jane or Beth were teaching a screenwriting class or a creative writing class, wouldn't this be just the type of feedback they would want? Are we much different than an analytical think-ing class or the creative writing or arts appreciation class? I don't think so.

In a 1990 essay, Brunsdon poses the question "What is good televi-sion?" Like most before her, she then declines to answer it, yet does take the question seriously, and offers some possible standards. And the ques-tion still nags at the edges of cultural studies. We know that all popular TV is not equal in quality, yet we cannot define what quality is, so we fall back on questions of popularity and ideology. What McGuigan calls the "drift into uncritical populism" (1998, 592) has the effect of flattening the aesthetic landscape, just as much as the elitist assumption that ev-erything on commercial networks is trash. This uncritical celebration of populism permits left-declaring scholars "an ideological veiling of the so-cial positions of... academic intellectuals, hindering a recognition of the political limitations, obligations, and possibilities inherent in these po-sitions" (Gripsrud 1998, 533)—and allowing them to conceal their own aesthetic preferences. They do not have to say openly that they don't like the genre they are celebrating for its liberatory potential, or if they do like it, they find all kinds of aesthetic qualities that actually do make it "good." Or if all else fails, they admit to liking "schlock," but only because they understand it "ironically." Williamson (1986), imagining a "radical left critique of *The Price is Right?*" asks, "with all our education, have we nothing more to say than 'people like it'"? (p. 19). What audience studies like mine confirm is perhaps obvious – *Dr. Quinn* is not *The Price is Right*, or *Jerry Springer*. TV genres like drama, comedy, and even soap opera may produce "bodies of work" that can be stored, studied, and reviewed using aesthetic criteria. Ephemeral reality shows, whose appeal depends on novelty and immediate emotional response, are very different, even if they may be legitimately analyzed for their sociological or psychological functions. As Kathy, the reflective "scholar-fan" mentioned in chapter 3, writes, "There is plenty that is bad in television, but well-drawn characters who nurture the imagination and show us the possibility in the development of our own higher nature and in what we can hope to find in others—that's not a negative thing." Dismissal of middlebrow shows as clichéd, sentimental schlock, often without watching them, is indeed elitist, depending heavily on easy social and gender-based stereotyping.

Of course, intrinsic aesthetic worth is a slippery concept. Aesthetic judgments are always culturally situated, as Bourdieu claims, and the same text may be appraised very differently depending on its cultural and historical context. In a study of the "meaning" of John Constable's highly popular painting "The Cornfield," Chaplin (1998) discusses how the picture is "edited" in its different contexts—first as an unconventional and unfavored work, then as a unique, "fine art" painting hanging in the National Gallery, and eventually as a mass-produced image on calendars, advertisements, and even wallpapers and firescreens. The image remains the same, but its status is completely altered. While Kellner (1999) argues that *The X Files* is a uniquely innovative and really "good" program, British and Australian *Dr. Who* fans believe it is just another example of American, special effects-based trash, drawing on a pervasive anti-American discourse: "The 'superficial' characteristics which are aesthetically bankrupt in this discourse are strongly associated with American texts" (McKee 2001). British people are often amused that long-running costume dramas like *Upstairs, Downstairs,* often dismissed as soap opera in Britain, are hailed as serious works of art in the United States, when "edited" as "Masterpiece Theatre" offerings on Public Broadcasting. And many *Dr. Quinn* fans have become dismayed that since being syndicated on the Hallmark Channel, the program has been redefined as a sentimental family drama. This goes beyond the issue of literal editing cuts to the more fundamental issue of "framing" the show; some old time DQMW-L listers decry the fact that new fans, who have seen it only on Hallmark, do not seem to understand it as the progressive, feminist text they grew to love.

However, a grasp of this important point does not mean that quality must forever be a forbidden concept in cultural studies. Popular audience studies, as well as just adding to the anthropological literature on cultural practices, can give us information about what *kinds* of texts have the potential to be understood and interpreted in a more thoughtful, aesthetic way. It is hard to make easy comparisons, since we do not always know exactly how research is carried out, or what questions are asked. However, studies like McKinley's of *Beverly Hills 90210* (1997) suggest to me that fans enjoy the program for all sorts of reasons, but that they apparently do not debate its "qualities" very much, while fan discourse of cult favorites like *The Prisoner* is centered on "fan-as-scholar" critical analysis. Maybe we could undertake a radical critique of the values embodied in *The Price is Right,* but I doubt there would be much point in thinking too carefully about the aesthetics of the show, any more than fans do.

This kind of study can also be invaluable in breaking through the invisible wall that still persists between scholar and fan, offering us an appreciation of the way that popular media are truly integrated into our culture. As scholars, we know it's perfectly possible to enjoy soap opera *and* grand opera, great symphony *and* great rock music. Yet we almost seem bemused when "they" do likewise, moving comfortably from genre to genre. As we become more comfortable with the complexity of aesthetic appraisal, it seems less important to debate whether a genre like television is "art." We may be able to approach the idea of what constitutes richer and more aesthetically rewarding television, without getting into absurdities such as whether *Dr. Quinn* is "better" than an Ibsen play. Maybe this can enable us to be braver about stepping out into the minefield of value judgment, and saying that yes, we should be able to advocate not just for more politically correct depictions, or opportunities for pleasurable subversion, but also for the "genuine widening of sensibility and emotional range" that Hall and Whannel so cautiously mentioned forty years ago.

6

CJ'S REVENGE
A Case Study of News as Cultural Narrative

On October 10, 1991, ABC's news magazine *Prime Time* ran a story titled "Angel of Death." It began with a montage of shots in bars, discos and other nightspots, showing people dancing and drinking, against a background of flashing neon signs advertising "topless" shows and other dubious pleasures. Over the throbbing dance beat, Diane Sawyer intoned:

> She is called the Black Widow. She says she has AIDS and she's bent on revenge. She says she's going to clubs and bars and sleeping with lots of men so that she won't be the only one to die. No one knows who she is, but she's the most talked about and most feared person in Dallas.

Reporter John Quinones then retold a story that had been causing a sensation in Dallas since early September, reaching the national press by the end of the month. It began when *Ebony* magazine published a letter in its "Advisor" column from a writer signing her/himself "CJ, Dallas, Texas." The entire text of the brief letter was: "I have AIDS. No one knows it. I go to clubs more now so I can meet new men. I feel that I am a beautiful person and I couldn't believe I got it. I sleep with four different men a week, some times more. I've slept with 48 men so far, some of them married. I feel if I have to die a horrible death I won't go alone. I know I'm not right in what I'm doing. Can you tell me what's wrong with me? Why don't I feel guilty?" (*Ebony* 1991, 90).

Prime Time reporters were told by the *Ebony* managing editor that the letter was not verified, but was printed "as a warning to readers."

The "Advisor" compared the letter writer to a serial killer, and advised him/her to seek treatment and "stop your vengeful actions," continuing that "the purpose of publishing your chilling letter is to warn those who are still in the habit of picking up strangers . . . that there are people like you on the loose who won't hesitate to do them in." The letter apparently struck a chord, leading to a flood of calls to a local African-American radio talk show host, Willis Johnson, who claimed to have received "the exact same letter" from a woman in 1989. Later, a woman had introduced herself to him in a Dallas nightclub, claiming to be CJ. He was completely convinced that her claims were real. Johnson issued a plea for CJ to call him, which she apparently did on August 31. He set up an interview with her on September 4. Extracts from the interview were played on *Prime Time,* the transcript running over a picture of a reel-to-reel tape slowly rolling, spliced with more darkly-lit shots of bars, where presumably unsuspecting men drank beer and danced, in slow motion. The recorded voice of CJ is interspersed with comments from Johnson:

Willis Johnson:	Do you feel guilty at all about . . .
CJ:	Doing what I'm doing?
Johnson:	Yeah.
CJ:	No, I don't. I wanted a marriage, I wanted children, but I can't do that now. I blame it on men. Period. Not just one man; he gave it to me, but I'm doing it to all the men because it was a man that gave it to me.
Johnson:	(Voice-over). On the other end of the phone was a devastating woman. She was cold, calculating, and this methodical about what she was doing.
Johnson:	Do you frequent all clubs?
CJ:	Yes. Especially Dallas. Men are so weak here in Dallas. Behind a pretty woman, a sexy body . . .
Johnson:	But CJ, realize how many lives you are affecting here. Men who have families and children and . . .
CJ:	Well, that's not my concern. That's their concern.
Johnson:	Do you feel anything at all?
CJ:	No, I don't.
Johnson:	Nothing?
CJ:	Nope. Revenge.

Prime Time went on to interview Assistant District Attorney George West, who had made it a personal crusade to find and arrest CJ. Police in Dallas had refused to get involved because no one had filed a complaint, but after media and public pressure, they too mounted a search for CJ on October 9.

Then on October 22, the Dallas police announced they had located the writer of the letter, a 15-year-old girl who did not have AIDS, but who was grieving for a friend who had died of it. The woman on the tape turned out to be the most prolific of about a half dozen women who had called Johnson's show and other local media. She was a 29-year-old woman, who also did not have AIDS, and whose motives were unclear.

As I watched the saga unfold, I was convinced that no actual CJ would ever be found or arrested. The reason? I saw the story of CJ as undoubtedly an urban legend run wild—a product of oral folk tradition that, because of its cultural salience, had become transformed into "news." This chapter is about how that happened, and how an essentially anthropological analysis of this case study may shed further light on our understanding of news as a form of cultural narrative that may enter the public consciousness. In this case, my entry point into exploring this dynamic is a story, or "text" (albeit a fluid one), rather than the "audience," as in the scandal discussion. Nevertheless, my perspective is similar, as I ask why certain narratives become salient, and how their salience is enmeshed in the fears, concerns, even pleasures of a particular cultural moment. This particular story shows the way producers/writers and audiences are inextricably linked to inspire, feed, and disseminate it widely and convincingly.

AIDS IN LEGEND

The folklore literature is full of AIDS jokes, legends, and other popular responses. AIDS was first diagnosed in 1981; as early as 1982, a story was circulating in San Francisco about a gay AIDS sufferer who would have sex indiscriminately and then show his lesions to his sex partner with glee, saying, "I've got gay cancer. I'm going to die, and so are you" (Smith 1990, 116). By 1986, as AIDS invaded the straight community, the most common form of the legend involved an infected woman. *USA Today* printed this report in October 1986:

> Novelist Jackie Collins . . . shared what she said was a true story on Joan Rivers' show. A married Hollywood husband picked up a beautiful woman at a bar. They enjoyed a night of passion at a good hotel; in the morning he rolled over to find a sweet thank you note. Class, he thought, real class. Then he walked

into the bathroom and found, scrawled on the mirror in lipstick: "Welcome to the wonderful world of AIDS." Not knowing if the woman had been kidding or not, he didn't dare have sex with his wife, and it could be a long time before he'd know whether he had become an AIDS victim. (*USA Today,* October 22, 1986, quoted in Goodwin 1989, 85)

Gary Alan Fine (1987) collected many similar versions, and Paul Smith documents the story in several European countries. It was printed in the British daily *Star* (February 25, 1987):

A British industrialist visiting Miami hired a lady of the streets to keep him company for the night on his stopover. The following morning when he woke, his companion had gone. On entering the bathroom . . . he found the following horrific message scrawled in lipstick on his mirror: "Welcome to the AIDS club." (Smith 1990, 114)

Smith shows that the story was circulating widely in Britain, Austria, and the United States during 1987. Newspapers and magazines were reporting it, sometimes as a legend, sometimes as an interesting, if unverified story, and word of mouth was doing the rest. He also points to the numerous antecedents to the AIDS story, in the form of similar lore about herpes, gonorrhea, and syphilis. The oldest example he cites is a tale told by Daniel Defoe in his 1722 *Journal of the Plague Year,* about the 1665 plague:

A poor, unhappy gentlewoman, a substantial citizen's wife, was (if the story be true) murdered by one of these creatures in Aldersgate Street. . . . He was going along the street, raving mad . . . and singing; the people only said he was drunk, but he himself said he had the plague upon him, which it seems, was true; and meeting this gentlewoman, he would kiss her. She was terribly frightened . . . and she ran from him, but the street being thin of people, there was nobody near enough to help her . . . He caught hold of her, and kissed her; and which was worst of all, when he had done, told her he had the plague, and why should she not have it as well as he? (Smith 1990, 131)

Similarly, Vietnam veterans told tales of prostitutes who deliberately infected GIs with "Black Rose," a supposedly incurable form of venereal disease (Gulzow and Mitchell 1980). This name is echoed in some version of the AIDS story, in which the victim receives black roses from the person spreading the disease.

The updated, AIDS version continued to circulate widely throughout the 1990s. Teaching in Minnesota through 1996, I received several variants from students each time I taught an introductory Folklore course. One version described the woman as "the Black Widow," the same term used in the *Prime Time* CJ story. In the 1990s, the Internet bulletin board FOLKLORE was reporting variants from almost all U.S. states, and by 2002 it has become one of the most thoroughly-documented legends on the popular urban legends website, snopes.com (Mikkelson 2000). In recent years it has faded somewhat, as AIDS has become culturally assimilated and has been supplanted by new horrors, such as terrorism, although my students in Florida continue to report versions regularly. We may assume that stories like these were also in wide circulation in Dallas, and the original letter writer was inspired by them. It may also be that her letter itself was an item of folklore, of the kind that circulates through xeroxed versions of jokes, parody memos, and letters—and now the Internet (Dundes and Pagter 1975). The talk show host's claim that he received an identical letter in 1989 could suggest the existence of an item of "xeroxlore."[1] Furthermore, according to Charles O'Neal, editor of the *Dallas Examiner,* "three or four years ago a similar letter made the rounds in Washington, D.C., and New York City" (Clemmons 1991, C5).

FROM LEGEND TO NEWS

But why did the newspapers and other media like *Prime Time* cover the Dallas story so extensively and, in the case of broadcast media, so uncritically? We tend to think of news and folklore as the opposite of each other; news is factual and verifiable, while legends are false and unverifiable. Recently, cultural researchers have sought to show that the line between news and legend is not so clear (e.g., Bird and Dardenne 1988; Bird 1992a; Lule 2001; Oring 1990; Smith 1992). News, like folklore and myth, is a cultural construction, a narrative that tells a story about things of importance or interest. Journalists like to think that news somehow mirrors reality, that it objectively describes events; news is "out there" to be discovered. But clearly news is not "out there." News does not exist until it is written, until it becomes a story, and what is deemed newsworthy owes as much to our cultural conceptions of what makes a "good story" as it does to ideas of importance or significance. Student journalists are encouraged by their textbooks to "find the story" in an event, using the same kind of criteria—conflict, drama, novelty—as any good oral storyteller.

As Bennett (1983) points out, the drive toward story tends to formularize news: "Any communication network based on stories will become biased toward particular themes" (p. 88). Thus reporters learn to find familiar stories in disparate events—the "passenger who missed the plane that went down," or narratives of government waste or inefficiency.[2] As Berkowitz (1997) shows, reporters are able to handle apparently unique, "what-a-story" events through the use of conventional narrative techniques that call on familiar themes. Perhaps this was shown most dramatically in the news media's response to the terrorist attacks of September 11, 2001, when all media, from tabloids to TV to "serious" newspapers offered similar stories of heroes, survivors, and villains (Bird 2002).

Most important in this context, news both reflects and reinforces particular cultural anxieties and concerns. It goes in waves; many scholars have demonstrated, for example, that waves of reporting about teenage suicide or child abuse do not necessarily reflect actual changes in the rates of these problems. In the United States, the summer of 2001 was dubbed "the summer of the shark," because of several highly-publicized shark attacks. In reality, there were no more attacks than usual, and sharks suddenly disappeared from the media after September 11. So news reflects waves of interest, and in turn feeds the anxieties that have produced the interest in the first place (e.g., Best 1990; 1991b).

Thus sometimes the media, without even knowing it, feed from rumors and folklore, restate these rumors, and the stories essentially become "truth." For example, it has become "fact" that every Halloween, children are injured or threatened by poisoned candy or razor blades in apples, and newspapers print dire warnings to parents. Yet scholars who have investigated this have found no actual cases of this happening, apart from one that involved a man poisoning his own son (Grider 1984).

These scares develop in part because of the very nature of news reporting. Notions of objectivity require that information be verified, but that verification frequently involves simply confirming that someone indeed said something. The quote itself becomes a "fact" when reported in the media. Thus, folklore about poisoned candy, children mutilated in mall bathrooms, or abducted by Satanic cults is shared, and passed on by everybody, including people who function as news sources (Victor 1993). Sources provide stories, and legend becomes news.

Michael Goss (1990) documents this process in assessing the case of the "Halifax slasher" in 1930s' Britain. Over several months, a man armed with a razor was reported to be attacking people at random in Halifax, Yorkshire. The case received wide media coverage, and people were quoted as having been attacked, or at least as having known someone who was. Eventually it was dismissed as a hoax. As Goss comments: "uncatchable 'super-criminals' always pose questions about how the 'system' is performing and the ways in which it has left us anxious and vulnerable" (p. 103).

Which brings us back to CJ, who did indeed become a sort of "supercriminal," as "ordinary people" were quoted as saying she must be found and prosecuted as a mass murderer. She certainly left people, particularly men who frequented singles bars, anxious and vulnerable. "Guys come in here to talk. First they joke – 'I just saw CJ out on the dance floor.' Then they get serious. They say, 'I don't mess with nobody. I'm going straight home,'" a nightclub men's room attendant told the *Dallas Morning News* (Precker 1991). KXAS-TV, a local station in Dallas/Fort Worth that covered the story extensively, frequently featured interviews with bar patrons who described how they had changed their ways and become more cautious. Other stories on the same station had experts analyzing CJ: "She doesn't have a positive self-esteem" (Sept. 9, 1991).[3] CJ's existence became a given, essentially because enough objective "facts" were being generated about her.

The feeling of cultural vulnerability was what the CJ scare was all about; to see it as simply a hoax on the media shows a one-dimensional view of news communication. In the 1990s, AIDS was widely perceived as a new plague; like syphilis during the Renaissance, it was seen as a "disease that (is) not only repulsive and retributive but collectively invasive" (Sontag 1989, 46). It is an epidemic that still causes widespread anxiety and fear, especially among those who fear their lifestyle could leave them vulnerable. Legends typically grow from anxiety and fear—maniacs on the loose, corporations that poison us, tanning booths that fry our insides. Media quite frequently see legends as exactly what they are, and disregard them; a *Dallas Morning News* reporter did make the connection between CJ and the AIDS legends, offering the hoax theory as one possible explanation for the story (Precker 1991).

But media are just part of our culture too, and reflect the concerns of the culture around them. Fringe tabloid newspapers, unconstrained by questions of objectivity and fact, know that and consciously feed from it. As a tabloid writer told me: "When looking for ideas for stories, it's good to look at fears, and it's good to look at real desires. That's why a lot of people win lotteries in the stories, and why people get buried alive all the time..." (Hogshire 1992). Mainstream media are more cautious and responsible, but they are still part of the cultural complex that saw AIDS as one of the greatest threats of all time. The vague, locally variable legends about the AIDS carrier were convincing enough to make many papers, because they did strike a resonant chord. The story "feels" true. *Ebony* Managing Editor Hans Massaquoi, explaining why the magazine printed the CJ letter without verifying it, said, "It sounded very believable" (Precker 1991).

And the media gave a specific name, face, and voice to the anonymous threat. In sociological terms, "labeling" functions as a way of defining a threat. Thus the media, by giving specific name-tags, bring the identity of mass murderers into sharp focus—Son of Sam, the Boston Strangler, the Halifax Slasher, the Yorkshire Ripper. "The mysterious, as yet uncaptured murderer is endowed with a recognizable image . . . a distinct label makes the faceless, anonymous peril seem less ambiguous" (Goss 1990, 105).

The women purporting to be CJ provided that label, and even provided a voice, personifying the anonymous peril of AIDS, and giving the media real quotes, real "facts" that turned a legend into news. "These labels are obviously useful to the media in presenting and codifying their news reports, but inasmuch as they bring a measure of compactness and form to the mystery assailant, they are equally useful to us, the audience" (Goss, 105). In effect, news stories that seem most credible to reporters and audiences alike develop their legitimacy through the construction of a body of "objective evidence," derived through a conventionalized process of newsgathering. The CJ story provided a wealth of such evidence that unfolded as the story grew. First we had the direct quotes, taken from the letter and audiotapes. A Dallas reporter recalls the impact of the audiotape: "That's what sold this thing, that it was such a convincing tape. . . . It was this good, sexy sound-bite that made all the difference" (Precker 1993). Then, the "person in the street" (or bar) was encouraged to comment angrily, fearfully, and colorfully, against a backdrop of visually striking stock footage of "the bar scene." As the story progressed, an artist's drawing was commissioned by the KXAS-TV station, using Willis Johnson's recollections of the "CJ" he had once met in a club, and this image was used everywhere, usually with total acceptance of CJ's reality: "A woman . . . is using the disease to kill. The woman is CJ, and this is what she looks like" (KXAS-TV, Sept. 4, 1991). As a newspaper reporter commented, "Every little story had that face looking at you—this attractive young woman was out there to get you. Adds to the drama" (Precker 1993). And to frame the story conventionally, the audience heard the opinions of police, District Attorneys, psychologists, and other "experts." All the pieces were in place to transform the vague, dispersed legends and rumors into the concrete, labeled reality of news.

THE CULTURAL SALIENCE OF CJ

This process would not have been so complete if the story had not already been culturally resonant. The AIDS/CJ story is a supremely relevant narrative for many audiences in several ways. For one thing, AIDS has been generally perceived as the most lethal plague in generations; it is a genuine threat. And many AIDS sufferers have spoken about their anger and bitterness at becoming a victim, and the consequent wish to lash out at society. The scenario is possible, and individuals have been prosecuted for knowingly exposing people to infection (e.g., *Dallas Times-Herald,* 1991, Gadsen, 1994). As Quam (1990) mentions, "In the case of AIDS, the culturally-defined line between sickness and crime is blurred" (p. 34). The AIDS sufferer is a victim-turned-villain in the popular mind, recalling the lepers and maimed, crippled evil-doers of popular culture.

Best (1991a) discusses the major role of "villains" such as mass murderers in folklore and popular culture, including news, talk shows, and other informational media. All these media emphasize that these evil figures are lurking, ready to strike anyone at any time: "Consider the villain who strikes at random, for no comprehensible reason. This is, of course, a very powerful, very frightening image. It suggests that everyone is equally at risk, that evil can strike suddenly, for no reason" (p. 113). The tale, then, strikes a chord that is equally resonant in oral traditions of story as it is in media definitions of story because both have the same cultural roots. We saw the same theme played out in the post-September 11 coverage of the hijackers, some of whom had been living among us in Florida and elsewhere (Bird 2002).

More specifically in the CJ stories, the AIDS villain is almost always a woman. This is inconsistent with the way AIDS is actually spread through heterosexual contact; while women are relatively likely to be infected by men, men have only a small chance of catching AIDS from a woman. Yet the woman perpetrator is perfectly consistent with wider cultural attitudes. In legend, the man she infects is usually married, or at least seeking illicit sex (in earlier variants the woman is often a prostitute). She is, then, the embodiment of the just reward—like many legends, she demonstrates that immoral behavior will be punished horribly and disproportionately. Grover (1988) points out that non-drug-using prostitutes have a lower HIV infection rate than other women, probably because they are more aware of the risks involved in their job. She argues that "prostitutes are taken as embodiments of infectiousness less for their actual risks and rates of infection than for their symbolic and historical status" (p. 26). Although CJ is not characterized as a prostitute, she is seen as a predator, an abnormally promiscuous woman. In fact, KXAS-TV ran a story that underlined the perception of CJ as even worse than a prostitute, interviewing a self-identified "hooker" named Maggie Patterson (Oct. 10,1991). Patterson's opinion: "Well, I think she's low-down. She should go to the doctor and get some help, because she's making it bad, you know, on other women that's out there selling their bodies and stuff, you know, a lot of women like me. I use plenty of rubbers and things, but she ought to turn herself in." The message is clear: Prostitutes are open about what they are doing, and men know what they are getting into, while women like CJ purport to be "ordinary" women, while harboring and spreading death among the unsuspecting.

Thus both the legend and the CJ story spread like wildfire among the singles bar and pick-up joints of Dallas, as a cautionary tale for *men,* rather than for women. CJ, like the woman in the legends, was a horrific siren for the AIDS age. She was characterized as unusually beautiful, seductive, and difficult to resist. Willis Johnson, describing the woman he later believed was CJ, said, "A man would have a difficult time resisting her" (Clemmons 1991, C5).

Taylor (1990) describes a poster featured in an AIDS campaign in the African nation of Rwanda. It depicts a woman saying "Come, let me give it to you," while a man runs away, crying, "Oh, no! You won't infect me with AIDS!" She is naked; he is wearing underwear. According to Taylor, this campaign "reflects the ambivalence of Rwandan cultural attitudes to female sexuality" (p. 61). While Rwandans do not deny women the right to sexual enjoyment, "There is . . . a contradictory tendency inherent in the culture both to encourage, and to fear, healthy female sexuality" (p. 63). It is not a stretch to find the same ambivalence in our culture; we use the sexually aggressive, always-ready woman as a pervasive dream figure, yet we fear the damage she might cause. CJ is a sexually-available dream figure whose polluted body turns the encounter into a nightmare. She is worse than clearly disfigured villains, because her pollution is invisible.

Treichler (1988) suggests several stages in the developing cultural awareness of AIDS. The folk legends of the disease closely mirror the changes she describes. During 1981–85, awareness was evolving slowly, and AIDS was clearly perceived as a gay disease that did not threaten the "general public." The folk legends tended to be restricted to stories about gays infecting each other. From July 1985 to December 1986, the death of Rock Hudson provided a "turning point in national consciousness" (Treichler, 196). Although Hudson was gay, his death opened the discussion of AIDS to suggest anyone could be infected, and that it was hard to tell who might be a likely "carrier." During the third period of development (1986–1987), AIDS was "perceived as a pandemic disease to which sexually active heterosexuals are vulnerable" (p. 196), and the heterosexual versions of the legend began to proliferate. Treichler describes how during late 1986–1987, "Major stories on AIDS as a threat to 'all of us' appeared, for example, in *Newsweek, US News and World Report, Time, Scientific American,* the *Atlantic,* and the *Village Voice*" (p. 212). As she points out, "It is interesting that by 1986, when women were more central to the AIDS story," scientists and physicians were speaking of "sexually active" males and "promiscuous females" (p. 213).

During the earlier days of AIDS, gays and bisexuals were assigned the role of "Contaminated Other": "The bisexual is characterized as demoniacally active, the carrier, the source of spread, the sexually insatiable" (Grover 1988, 21). As awareness of heterosexual infection grew, prostitutes were assigned that role, as they had been in other plagues, and the earlier AIDS legends feature prostitutes as carriers. Later, the labeling extended to any sexually active woman, as reflected in the legends, including the particular legend variant that became the CJ story. As Treichler notes, in the early 1980s, women were generally regarded as particularly well-protected against AIDS, but gradually this changed, until women were perhaps the most suspect of all: "Like the virus, wearing an innocent disguise, are we not double agents, in league with the enemy?" (p. 221). Once again, we see a precursor to that notion of woman as "double agent" in the iconography of syphilis and other sexually transmitted diseases. As Gilman (1988) writes, "Already in the Middle Ages, woman was shown as both seductive and physically corrupt—for example in the sculpture of Madam World from the St. Sebaldus Church in Nuremburg. Female beauty only serves as a mask for corruption and death, demonstrated by the contrast between the front and the back of the sculpture" (pp. 95–96). He describes a gradual shift from the notion of a male "victim" to that of a female "source of corruption," pointing to such images as a title-page illustration for a 19th-century text on syphilis, that shows a syphilitic woman hiding her ugliness behind a mask as she attempts to seduce a male victim.

Thus a woman "carrier" is more likely to strike fear into the general populace. Emke (1992) analyzes how AIDS "carriers" became defined as "folk devils" (Cohen 1980) in Canadian culture, as "carriers were seen to cross a boundary that separated the homosexual closet from the heterosexual world" (p. 15). As long as AIDS could be dismissed as a "gay disease," the "moral panic" about AIDS was relatively muted, and gay victims of the disease could be stigmatized as guilty and deserving of their fate, in contrast to "innocent" victims. The early 1980s legends of the gay man who deliberately infected others had little salience outside the gay community. Likewise, as long as the media characterized AIDS as confined to gays, widespread perception of a universal threat was absent. Altman (1986) quotes a journalist speaking at an AIDS forum: "It would be dishonest not to say we couldn't sell the AIDS story early on because it was about gays" (p. 16). Only when children, straight men, and women became victims was the stage set for general "moral panic." The Dallas

TV coverage of the CJ scare illustrated that moral panic clearly. An early story discussed reaction to the CJ letter, which was posted in the bathrooms of many Dallas clubs. "Experts say folks need to know that AIDS is no longer, or never was, a white gay man's disease...Doctors agree, saying the disease has spread to the heterosexual society, and no one is safe" (KXAS-TV, August 30, 1991). The story ends with the reporter speaking over the familiar dark dance-bar scene: "It's Friday night, it's the Dallas club scene. Tonight folks are having fun. But tonight, folks are thinking about a deadly disease that could end their lives."

A woman, and particularly a beautiful, hard-to-resist woman, is especially evil, deceitfully carrying the virus out of the closet (where it belongs) and into the world of "normal" men. Significantly, the original letter to *Ebony* never identified CJ as a woman, only as "a person." If CJ had been perceived as a man, who was infecting other gay men, the furor would never have developed. Quickly, she became known as a woman, in keeping with the time period's folk legends. As Sontag (1989) points out, although "at-risk," and therefore already stigmatized groups are terrifying in the role of carrier, in many ways the "innocent" are more frightening, since they "potentially represent a greater threat because, unlike the already stigmatized, they are not easy to identify" (p. 27). Beautiful women were not perceived as a stigmatized group. The "moral panic" surrounding CJ was so great because she was seen to *masquerade* as innocent, using this mask to cross the boundaries and infect the truly innocent.

Furthermore, while the CJ story struck fear into everyone in Dallas, it particularly affected the African-American community. CJ was not just a beautiful woman; she was a beautiful Black woman. Worth (1990) points to "the 'colonization' of values concerning appearance (such as seeing lighter women with straight hair as being more sexually desirable)," (p. 122), and CJ's image was a perfect example of "colonized" notions of Black beauty: She was described as light-skinned, straight-haired, and slender. This apparently made her especially threatening. A woman active in the local African-American community commented, "Men have this idea that women have a power they can't resist. And if they have this power and have AIDS and he sees her, he's out of control. I think they need to put a little padlock on their zippers" (Lyons & Needham 1991, A8).

Television news coverage of the story did not make an issue of race verbally, but visually it tended to reinforce stereotypes of African-American sexuality. Stories repeatedly used the same footage of Black people dancing in dimly-lit nightclubs, apparently disregarding the beautiful threat lurking among them. But the stereotypical beauty of CJ allowed her image to be "mainstreamed" beyond the African-American community, since she fit so perfectly the age-old image of the promiscuous, exotic, "dusky" woman who lured white men into sin.[4] When the CJ story began, accounts suggested that she was targeting African-American men in particular, but this aspect quickly disappeared, so that by the time *Prime Time* covered it, the familiar club scenes included dancers of all races.

Meanwhile, for women the tale may function as a revenge fantasy. Fine (1987) suggests that women may find the story attractive because it pays men back for abuse or rape. Indeed, that seems to have been the motivation of the main caller who claimed to be CJ. She expressed her anger and hatred of all men, and her desire to wipe them all out. If she was looking to scare men, she was clearly successful, as we can see from the response of the CJ crusader, Assistant District Attorney West, who even after the story died expressed his concern: "I'm sure that the anger of someone who has been infected and is then lashing out at others is absolutely real" (Precker 1991, 3C).

Prime Time uses West to make its most striking visual statement about the sinister and predatory evil that lurks "out there" in the midst of an innocent, unsuspecting community. Throughout the piece, there are numerous views of dark dance halls and flashing lights, conjuring up a picture of a seamy underside to the city where "innocent" people go at their own risk. The story finishes with a shot of West's back as he looks out of the windowed wall of his office onto the Dallas skyline. The camera pans back to show him in silhouette, framed by distinctly church-like windows. As the camera lingers, we hear West's voice: "I look at a beautiful community, and I know that there's a walking plague out there. CJ is out there. She could be anywhere."

As we continue to watch West look out over the brightly-lit, "innocent" Dallas skyline, we hear CJ's voice again: "I'm doing it to all the men, because I'm very angry, very angry." Finally, we return to a grim-faced Diane Sawyer, who concludes: "And as of tonight, the Dallas DA's office and Police continue to follow leads, but so far have made no arrests."

Within days of the piece airing, the mother of the girl who had written the *Ebony* letter went to the police, believing that events had gotten out of hand. Following this, Dallas Deputy Police Chief Ray Hawkins told the *New York Times:* "Our attitude is that there is no CJ as purported to be. However, there are lots of CJ s out there, either knowingly or unintentionally spreading HIV" (*New York Times* 1991, A14). In Texas, such an action is a felony punishable by up to 10 years in prison. At the time, Dallas had a high incidence of AIDS, ranking 12th nationally, with 3,200 cases diagnosed by September 1, 1991. During the height of the CJ scare, AIDS awareness grew dramatically, with activists handing out CJ t-shirts, leaflets, and condoms. The AIDS hotline received 1,200 calls in three days, more than in the previous month. The hotline added a new tape to its computer menu, called "CJ," which callers could access for counseling and advice (Krepcho, et al. 1993). According to the local health department director, "I look at what happened with CJ as a fire drill, something that has made people aware of danger and risk" (*New York Times* 1991, A14).

For 200 years, the media have been subject to hoaxes, as Fedler (1989) amply demonstrates. Frequently, these are a result of deliberate misinformation campaigns, and some people, such as the infamous Joey Skaggs, have made a career out of duping the media (Tamarkin 1993). Obviously, hoaxes are most successful when they seem plausible because they speak to something current in the culture. A noted example was the hoax launched in 1992 by the "Coalition against Fantasia's Exhibition," which held press conferences protesting the revival of Walt Disney's *Fantasia*. Among other things, the group complained that the "dancing hippos" scene demeaned overweight people. They were soon joined by other groups, protesting scenes like the "dancing poppies," which apparently glorified drug use. The coalition announced in late March that the campaign was designed to be an April Fools' hoax, but by then it had received coverage in local media, as well as in *Time* and the *Washington Post* (Tamarkin 1993). Twenty years ago, it is likely that the media would have dismissed this campaign as absolutely ludicrous and unbelievable, yet in the prevailing climate it was taken seriously. There seemed to be no more reason to discredit the Coalition as a valid news source than to discredit the legitimacy of government spokespeople.

The CJ incident was not a hoax, in the sense of a deliberate misinformation campaign. Yet like true hoaxes, it depended for its vitality on the cultural climate being right, thus legitimizing the sources of the story. And because the story reached so deeply into cultural fears and stereotypes, it took on a life of its own. Essentially, it was the coming together of anonymous rumors, tales, and legends, initiated by one individual, but fed by oral tradition and media alike. George Bernard Shaw once commented that it took the media blitz surrounding the Jack the Ripper murders to draw attention to the terrible conditions faced by the poor women of Whitechapel who became prostitutes and potential murder victims: "While we conventional Social Democrats were wasting time on education, agitation and organization, some independent genius has taken the matter in hand" (quoted in Goss 1990, 106). The 15-year-old girl who wrote the letter said she did it not to deceive, but to increase awareness of the deadly threat of AIDS. With the help of folklore and the media, she apparently did just that.

CONCLUSION: NEWS AS MYTH

Traditional accounts of news tended to see it as a product of sociological or historical forces. More recently, the notion of news as storytelling has become increasingly conventional, as part of an interdisciplinary reevaluation of storytelling. Much of this reevaluation has focused on the structural aspects of narrative–for example, how journalists construct stories in particular, stereotyped ways. This has opened up the study of journalistic practice immeasurably. However, a neglected area has been an exploration of how news narratives interrelate with the stories and concerns of the wider culture at specific moments in time. For instance Lule (2001) argues for an understanding of news as myth, and makes a good case for journalists' role as "scribes," analogous to ancient storytellers and bards. Like Lawrence and Jewett (2002), he shows convincingly how the archetypal myths of the hero are played out in news stories, couched in familiar and comfortable formulae.

However, Lule's application of "myth" is limited because of its dependence on Eliade's (e.g., 1958) and Campbell's (e.g., 1949) emphasis on archetypal or universal textual themes, with a nod to Malinowski's more functional, anthropological interpretation (1954). The weakness of this "universalist" approach is that it pays scant attention to the differences in time and place that produce particular cultural moments and narratives, rooted in particular histories. Furthermore, in focusing on texts, it ignores the participatory role of the culture in which the myth is embedded. It acknowledges the kinship of folk and print culture, but does not fully grasp the fluid, processual nature of that kinship. For Lule, as for many others who have analyzed news as mythical narrative, the focus is on the journalist, who is given the burden of defining the myth. Thus, "the language of news is what matters" (Lule 2001, 5).

Strangely, Lule argues that "rather than plumb the depths of the human mind, myth stays on the surface of the printed page" (p. 146). This statement seems to fly in the face of a truly anthropological understanding of myth more as process than text, and as a joint product of storyteller and audience. For instance, the "trickster" myths invoked by Lule in analyzing news coverage of Mike Tyson's rape case are most usually told not by a "scribe" or venerated authority, but by parents to children, in a participatory process that involves comment, discussion, laughter, constant revision, with rarely a "printed page" in sight (Trejo 1979). So while Lule suggests that the news audience—the cultural context in which the news exists—is almost irrelevant to the role of news as myth or narrative, I would suggest that it is crucial. It is important to think about what people do with the stories journalists tell—and to explore the way journalists' stories are embedded in their understanding of cultural concerns at any given moment. The CJ story certainly draws on numerous archetypical themes that have surfaced throughout history, as I have demonstrated, but its impact as a blockbuster story was just as significantly linked to the immediate circumstances of the early 1990s; it did active cultural work. Nightingale (1996) rightly argues for a more thoroughly anthropological understanding of this fluid participatory quality of myth, pointing out that audience-focused studies "in this sense . . . represent a functionalist incursion into the structuralist territory of the text" (p. 107). To focus only on unchanging archetypes is to neglect the dynamic nature of culture; Hall (1981) describes constant cultural transformation as "the active work on existing traditions and activities, their active reworking so that they come out a different way: they appear to 'persist'—yet, from one period to another, they come to stand in a different relation to the ways working people live and the ways they define their relations to each other, to 'the others' and to the conditions of life" (p. 228). Certainly, journalists set the news agenda, and they undoubtedly may function as authoritative scribes, but they are not outside the daily rhythms of culture any more than are traditional myth-makers and storytellers.

Analysis of cases like the CJ scare, while not constituting "audience study" as such, can show us that salient news stories may emerge from a swirling cultural cauldron of tales, rumors, jokes, fears, beliefs, and other narratives of semi-narratives. Through the techniques of news story construction, rumors crystallize into stories. As Gritti (1994) comments, "Rumor is surrounded by a halo of imprecision, while *faits divers* [human interest news story] builds upon precise questions: who?, with whom?, what? how?, why? etc." (p. 5).

Journalism is driven by a need to explain, and AIDS is something that is, in many respects, inexplicable. The legends frequently leave it unexplained, while journalism seeks answers, often through stories that personalize the vague, anonymous forces "out there." Carey writes, "Explanation in American journalism is a kind of long distance mind reading in which the journalist elucidates the motives, intentions, purposes, and hidden agendas which guide individuals in their actions" (1986, 180). Individuals and names make new stories, as CJ shows us. The *Ebony* letter writer gave the anonymous threat a name, the talk show caller gave her a voice, and the local news gave her a face, all of which the other media enthusiastically disseminated. The story did not die, even after the emergence of the "hoax." Indeed an interesting echo of the CJ story hit media in the United Kingdom and Ireland in September, 1995. A Roman Catholic priest in a small town in County Waterford, Ireland, claimed in a sermon that he had talked with a young woman who had deliberately infected up to 80 men in the area with HIV. A woman claiming to be the girlfriend of one of the unnamed victims told the British *Daily Express* that she had confronted the woman, who told her, "I hate men, I hate all men" (Sept. 14, 1995, p. 6). Father Michael Kennedy's avenging angel was described as "a tattooed woman on holiday from England." It's a very different image from that of CJ, but it is equally effective as a locally-salient representation of an exotic, worldly, urban demon preying on the inexperienced innocence of young, rural Irishmen. "I believe she is going to hell, because that is what she deserves," cried the wronged girlfriend. After about a week of heavy news coverage in British and Irish print and electronic media, the story died down. More serious newspapers pointed out the unlikelihood of a woman infecting even one man so easily, let alone 80, while neither father Kennedy nor anyone else was able to identify the woman or any of her victims. The inevitable conclusion is that Kennedy wove the familiar floating legends into a parable that, with the help of the media, spiraled out of control.

Recently, we saw the same process in action with the international circulation of a story that expressed not fear, but desperate hope. In the aftermath of the September 11 attacks on the World Trade Center, a story circulated around the world about a man on the 83rd floor of the 110-storey south tower, who survived the building's collapse by "riding the rubble" to the ground. Beginning as a rather vague rumor among rescue workers, it solidified, as reporters first identified the man as a firefighter, later naming him as Sgt. John McLaughlin, a New York Port Authority police officer and father of four. *Newsweek* confirmed the story, quoting rescuers and a doctor at the trauma center. Only later did it gradually emerge in the media that McLaughlin did not exist, and that it would have been impossible for such a thing to happen. The story, like that of CJ, was not so much a hoax as a folk narrative that grew from a rumor and took on a life of its own. "In addition to giving the public and the rescue workers hope, it gave the world's media a target for one of the most astonishing survival stories ever" (Alderson 2001). We see a similar process in action with any number of stories that become "fact" simply by virtue of being reported. Murad (2002) uses the example of a story fed to reporters by the Republican National Committee, which claimed that then-presidential-candidate Al Gore had ordered the wasteful opening of a dam on a river in order to create a photo opportunity showing him kayaking. The truth was that the dam was opened every day, but "reporters on the bus...decided it was too delicious to ignore" (Murad 2002, 89), and the story was widely reported. It was "delicious," of course, because it created a picture of Gore as a hypocrite. It took hold not because of any real newsworthiness, but because it provided a very specific anecdote that crystallized a widespread unease about Gore's honesty and consistency, and would become the kind of entertaining story that circulates in the everyday gossip and rumor that surrounds news.

Thus, we can see news not as some entirely distinct form of text. Although it clearly has its own distinctive generic properties, it is merely one strand in the complex web of culture. News itself is a complex genre; its mythological or storytelling function is only one dimension, along with the more conventional function of setting an agenda and chronicling major events (Bird and Dardenne 1988). But as Carey (1986) writes, "In fact, journalism can present a coherent narrative only if it is rooted in a social and political ideology, an ideology that gives a consistent focus or narrative line to events . . . " (p. 161). Sociology, history, and narrative theory have all helped to increase our understanding of the process of news. In my analysis of the CJ story, I have tried to draw more on the holistic, cultural focus of anthropology, reaching out from the story itself toward a set of connections between it and notions simmering in the culture at large. In this respect, although the analysis starts with the text, I believe such a "thick," contextual exploration also sheds light on the relationship between text and reception. As Abercrombie and Longhurst (1998) point out, such text-originated studies can contribute significantly to our understanding of media reception only if they are "further related to the place of the text in everyday life" (p. 161).

In exploring the relationship between folk narratives and media culture, I do not wish to argue that the two are identical. There are still areas of culture that are largely oral and owe little to media, and there are mediated narratives that owe more to a highly literate tradition than to folk culture. The relationship between journalists and audiences is not the same as that between traditional storytellers and their audiences, who interact face to face. In solely oral cultures everyone potentially is a storyteller, even though many in practice are not, while in our culture the journalist or other media producer inscribes the text, no matter how much input readers have. As Siikala (1984, 152) notes, mediated narratives have replaced folk narratives in most people's lives. Nevertheless, "Even though tales are produced commercially, people are still left with scope for interpretation. Nowadays, instead of the narrator, we increasingly come across the commentator expressing opinions on the narratives transmitted by the media." (p. 152). And, as I have tried to show in chapter 2, the narratives themselves become transformed and recreated in the hands of the public. Thus, like Oring (1990), I would argue for "the conceptualization of a single ideological domain to which folklore, news, literature and court cases belong . . . ." (p. 170). News as a genre is indeed distinct, but unless we start to see those connections, we will find it harder to understand why particular stories have salience at particular moments. "Big stories," "moral panics," and even tenacious "little stories" (like the Gore kayaking one) are not created by journalists alone; rather, journalists may also be the brokers for the stories a culture is already telling.

7

MEDIA ETHNOGRAPHY
An Interdisciplinary Future

INTRODUCTION: TAKING STOCK

As I come to the end of this book, and try to reflect on "what it all means," I have two people in mind. One is a young boy, a fourth grader I met while studying the impact of the introduction of computers into family life (Bird and Jorgenson 2002). When my research partner and I first entered his small trailer home in rural Florida, he was sprawled in front of the television, which continued to play throughout our interview with him and his mother. "Kevin" was an obese child; his breathing was labored, and his skin was pale and pasty. We talked about the computer and about other technology in the home, learning that he had his own TV/VCR in his room. Our interview was during the summer vacation, and we discussed how he spent those long, free days. "Watching TV," he told us. He wakes up in the morning, turns on his TV, and watches all day, often eating on his bed. He likes all cartoons, but he'll watch "most anything," and usually drops off to sleep with the TV on. His mother, busy with a young baby, is happy that he keeps occupied; she disapproves of the neighborhood children, and doesn't like him playing outside, where he may get into trouble. She feels "safe" when he's in his room, where she can keep an eye on him.

The other person is someone I've never met, except online, where we have had several stimulating personal "conversations." One of the *Dr. Quinn* fans, "Kathy," a middle-aged White woman, generally watches little television otherwise, but lives a very active outdoor life, reads voraciously, and has numerous hobbies. She has a special interest in Western and women's history, and her other main online group is an e-mail list for high-IQ people. She is a self-reflexive fan, who not only writes fan fiction and careful, critical analyses of TV texts, but has also done informal research of her own on what being a fan is all about, surveying members of her online communities.

Kevin and Kathy seem like mirror-images of the "bad/good" poles of living in a media world. By no stretch of the imagination could Kevin be seen as an "active" or "resistant" media user, even allowing for his young age. He cannot describe a show he has just watched in any clear terms; in fact, he really doesn't discriminate among different programs. He literally "goes with the flow" of his television, with occasional decisions to watch a movie, play a video game, or (rarely) use the computer. He cannot imagine what life would be like without his TV. Kathy, on the other hand, has used her media enthusiasm to learn, communicate with people, and enrich her life in many respects. She seems to epitomize the "active" viewer, making constant choices and remaking her favorite program both through fiction and analysis. Strangely enough, it would probably be Kathy who many would dismiss as slightly "odd" because of her fan "obsession," while Kevin's life might seem unremarkable.

THE REALITY OF THE ACTIVE AUDIENCE

As I think about these two, it strikes me that audience research has been successful in telling us a great deal about the Kathys, and really very little about the Kevins—audience research has become a very optimistic tradition, and much of my work belongs in that tradition. We now have quite rich documentation of audiences and their creative interaction not only with TV, but with all aspects of popular media. Media bring entertainment, diversion, food for thought, and shared symbols that unite people, whether they are connecting interpersonally or electronically. This is true in other cultures too; Kottak (1990) for instance, suggests that once television comes to small villages in Brazil, it actually opens people's horizons, allowing villagers to become aware of worlds outside their immediate vicinity, and decisions made elsewhere that affect their lives. It fosters a sense of national, rather than local identity. The burgeoning literature on fans (including my own work) suggests that shared fandom can be an immensely positive and energizing experience, which far from isolating people, may create new friendships and a sense of community.

Young people are apparently particularly creative in the way media are integrated into their lives. Willis (1998) convincingly describes the "symbolic work" that young people do, as they choose how to create their own identity through media:

> Repeated listenings to records and tapes, writing words down, memorizing them helps to make them personal and meaningful possessions. This is the practical means of establishing psychic if not legal possession: "music gives me strength"; "reggae is heart music, it gets me there"; a young white man using the notion of "sufferation" from Bob Marley to resolve and explain his own experience. (Willis 1998, 167)

Similarly, Drotner (2000) argues that in equally media-saturated Denmark, young people "are not only the social group to make the fullest use of the most media in reception terms, they are also the most active in media production" (p. 156), as increasingly sophisticated technologies "come to serve quite directly as building blocks in personal processes of digital production" (p. 157). Fisherkeller (1997) points to the diffuse way media are integrated into young people's lives, melding with education, family, and personal communications to shape imagination and identity. These and many other studies like them paint a picture of a generation that is seamlesslessly but actively integrated with all forms of media, so that "it makes little sense to analyze youthful media cultures as part of a secondary socialization that is distinct from interpersonal forms of communication in primary socialization" (Drotner 2000, 159). While Willis tends toward a celebration of this new reality, Drotner and Fisherkeller seem a little more cautious; they seem to be suggesting that whether we like it or not, we had better come to terms with this mediated cultural reality. Nevertheless, they are essentially optimistic in their enthusiastic discussion of media's role in their informants' lives.

In many respects, I share this enthusiasm. My own research has not focused on "cultural dopes," but has found active people who enjoy their interactions with media, even when they find media wanting, as many of my American Indian participants did. At the same time, I believe we do need to more fully interrogate the reality of living in the mediated culture that some celebrate, some fear, and some (including some audience scholars), feel nervously ambivalent about. Just as we have tried to avoid judgments about whether popular media are "good" in terms of quality, I believe we cultural studies scholars have also often felt profoundly uneasy about whether media are "good for us" in a more general sense. As Alasuutari notes, in both private and public discourse, "Media use is in many ways a moral question" (1999, 11). The "active audience" movement arose in large part to counteract the "cultural dope" view of media consumption, and it has been very successful in reconceptualizing the audience (us) as participants in media culture, rather than its victims. Yet many of us are still ambivalent. As scholars we wrestle with debates about whether audiences are really "active," and what that activity actually means in cultural and political terms. As teachers, we work to understand our students' media-drenched perspectives, while complaining to each other about their short attention spans and lack of wider cultural literacy. As parents, we count our children's hours in front of screens, and worry that their imaginations are atrophying—only to be amazed when we see them playing an elaborate game that incorporates everyday props, inventive role-playing, and media characters, as they interact with friends who share the same symbolic world. We wonder if media are benign "symbolic equipment for living" (Burke 1966) or insidious tools of economic, cultural, and political oppression. Or both.

Moores (1993) writes that we need ethnographically-oriented studies not so much for their own sake, but because they will help us reach a better theoretical understanding of "constrained cultural creativity and a proper method of interpretive cultural study" (p. 140). I believe this neat phrase—"constrained cultural creativity"—nicely sums up the direction audience research must continue to take. Yes, as audiences we are (or can be) creative, taking images and ideas provided by the media, and doing many unexpected things with them. Yet we are also constrained by the boundaries placed around the meaning of those media products, and by the choices that are actually available to us. We must come to grips with the issues of power and control that often seem to play around the edges of audience research. What are the implications of having media framing our everyday lives? Abercrombie and Longhurst (1998) comment casually about the way our lives intersect with images: "In day-dreaming about a forthcoming wedding, for example, the content of the day-dream will be informed by countless elements taken from media performances..." (p. 118). This sentence evokes the pleasure of the media-framed daydream; we imagine the excited bride with visions of her favorite TV wedding playing in her head. But let us not forget that this is a daydream that fuels a multi-million dollar wedding industry, an industry that encompasses everything from movie-inspired settings to elaborate photographs that look like glossy magazine ads (Lewis 1997), and that has driven the average cost of a U.S. wedding to $20,000 (Lamas 2002).

Unspoken is the pain—for those who share all these media-framed dreams but will never realize them, because they are too poor, too heavy, too "ugly." For instance, few people now deny the relationship between impossible, media-defined standards of female beauty and eating dis-orders or poor self-image among women (e.g., Stice et al. 2000). The most recent studies suggest that these problems now occur in younger children than ever before, and that they are becoming common among boys as well as girls (Robinson et al. 2001). The dramatic power of the media to define desirability and beauty is suggested in an anthropological study of the cultural impact of the recent introduction of television into a Fijian culture (Becker et al. 2002). The authors paint a dispassionately scientific, but heartbreakingly sad picture of changes effected in a matter of a year or two by the arrival of images of blonde, thin, imported beauties. In a society that traditionally valued generous female proportions, young girls quickly learned disgust for their bodies, discovered the binge/purge syndrome, and told interviewers: "I want their body, I want their size. I want to be in the same position as they are ... We have to have those thin, slim bodies ... " (Becker et al. 2002, 513). Certainly, these young girls derived great pleasure from this new medium. But we cannot ignore the negative impact of globalizing economic forces that ensure that Fijian girls' first media experiences are of *Baywatch* and glamorous soap operas. While there is certainly much evidence that local cultures are considerably more resistant and creative against the power of the globalizing media than once feared, nevertheless, in a culture that is not media-saturated and media-savvy, the impact of images can be especially powerful (Becker et al. 2002).

THE CONSTRAINTS ON PLEASURE

In this chapter, then, I want to explore some of the many sides of media reception—not with the intention of dismissing the important concept of the active audience, but in an attempt to examine the larger context in which the audience functions, and to acknowledge the consequences, positive and negative, of living in a media world. We may have moved on from simplistic cause-and-effect analyses of "media effects," but that does not mean the media are not powerful agents in the formation of our culture. The ideal of individual autonomy is so ingrained in Western culture that the concept of audience power has an instant, almost insidious appeal. Yet as Grossberg (1997) writes, "We need not the denial of strong media effects, but the careful consideration of how and where the media are effective, and how their effectivity is produced" (p. 341). While the active audience tradition has taught us that we, the audience, have a significant role in creating and operationalizing that effectivity, we cannot pretend that the power of corporate media producers can somehow be vaporized by the magic wand of audience creativity. For instance, my work on racial representation, like much other research, points to the way minorities, whether defined by ethnicity, gender, sexual orientation, or such less-considered categories as age or religion, frequently feel alienated and marginalized in mainstream media culture. They have to "work" harder to share in the pleasures of the active audience, since mainstream media mirrors dreams and ideals they do not necessarily share. I well recall chatting with some of the American Indian students who had participated in my first study of representation (Bird 1996), and who had come to hear me present the findings at a local conference. They enjoyed hearing their views represented, and began spontaneously and quite seriously discussing how we might all go to Hollywood and use the work to convince the powers-that-be to incorporate Native perspectives in programming. Then they laughed and admitted the reality that no one in power has any reason to listen to such thoughts, and that a professor was no more likely to be heard than a few Minnesota Indians.

There's no doubt: Studying active audiences is fun. I have enjoyed exploring the pleasures of TV fandom and everyday news conversations, as much as many other researchers seem to have enjoyed their discoveries of audiences making meanings. But I have found much less pleasure when it comes to working with people whose lives are considerably more constrained. Let's consider "Kevin," who seemed to epitomize the couch-potato-in-training. As we left his home, my research partner and I incoherently and guiltily discussed our feelings of distaste and discomfort with his family's bounded lifestyle. He lived in the country but never went outside; he lived in year-round sunshine, but his skin was pale and unhealthy-looking. He had been furnished with a computer to enrich his education, but he showed little interest, preferring 24-hour immersion in television. His young mother, tired from coping with him and a baby, talked about her fears for the future; she is worried about what will happen when her high-school-dropout husband's bad back finally stops him from working the manual labor jobs that support the family. She wants better for her children.

My partner and I were working in rural Florida with a very specific purpose. In cooperation with a local elementary school, we were studying a program its teachers had developed that used federal funds earmarked for low-income schools to buy computers and place them in "at-risk" children's homes (Bird and Jorgenson 2002). Their goal was to work with individual children using Internet-based lessons to improve reading and math, with the simple aim of raising their test scores and grades. Children had to commit to doing the lessons for 30 minutes a day; in practice, parents had to supervise and enforce this commitment. In return, the family was loaned an iMac computer, with the only financial obligation being the monthly Internet connection fee. The school hoped to help families become "computer-literate," offering training sessions, workshops, and in-home set-up sessions. My partner and I conducted ethnographic interviews with families, with two aims: To help the school staff assess whether the program was "working"; and to study whether and how the new technology was being integrated into the families' everyday life. As we learned more about each family, I began mentally comparing their experiences with the technology-savvy young people celebrated by Drotner (2000), Willis (1998), and others, who creatively worked across media to make and remake their culture. Grossberg (1992), in considering the celebratory work on media experiences, points out that the pleasure of enjoying "everyday life" is very much dependent on having the resources for leisure and consumption. "In everyday life, one has the luxury of investing in the mundane and the trivial, in the consumption of life itself. To offer the simplest example, there is real security and pleasure in knowing when and where and exactly for what (including brands) one will go shopping next" (1992, 149). He contrasts pleasurable "everyday life" with the "daily life" experienced by those without the resources to participate in the media world.

This is exactly the issue that we tried to explore in our study of computers in family life. We found that families were united in their belief that computers offered a veritable cornucopia of riches for their children; they talked of all the information on the Internet, and the skills that would open opportunities for the future. They mentioned all the things one could do with a computer, from games to multi-media applications. Yet for most, these opportunities still seemed unreachable. There were the practical problems of living in rural areas—unreliable phone lines that crashed, lightening strikes that knocked out the computer, unavailability (and expense) of broadband connections. As the Internet becomes increasingly a site of pleasurable consumption, it holds less appeal for people who have no credit cards, who cannot afford to travel, and must be careful with every dollar earned. Learning how to use the Internet, let alone related multi-media possibilities, requires a level of educational and cultural capital that is largely unobtainable for the poor (and these participants were by no means the poorest of the poor). We found that television was regarded as an essential in every family we visited, and it was enjoyed continuously and passively, as a necessary palliative at the end of a long, physically demanding day, as a babysitter for kids, and perhaps as a provider of dreams. One mother told us that she was trying to interest her husband in the computer (fathers were rarely engaged with the technology), by pointing him to the Web site for the TV show *Survivor*. Was he a fan of the show? Not exactly, but he dreamed of somehow getting on the program as a way to win money and "succeed," and so had looked at the application process. He never applied, and of course he would never have been chosen if he had. Self-described "rednecks" like him get their few minutes of fame being ridiculed on *Jerry Springer*, not modeling bandanas on *Survivor*.

So in our scholarship, we must come to grips with the fact that audience activity and creativity is not always an option. As Hardt (1998) writes, American media studies (and I would certainly include audience studies here) have long failed to incorporate an adequate consideration of class, with its entrenched social and cultural inequalities. "Communication studies remains informed by progressive notions of change that adhere to ideological and political constructions of a social reality in the context of middle-class expectations" (Hardt 1998, 51). Our work with these families juxtaposed the progressive ideology of technological innovation with the lived reality of people whose lack of cultural and economic capital placed huge barriers between them and creative, active engagement with the new technology. I also came face-to-face with this reality in my work on supermarket tabloid readers (Bird 1992a). Most certainly, I found playful, engaged readers who worked with the texts and integrated them into their lives. I found ironic, subversive readers, who enjoyed the parodic excesses of the tabloids, as celebrated by Glynn (2000). But I also found readers who quite literally believed in the most outlandish possibilities, from fossilized dinosaur eggs to alien abductions. Later I spoke with a former tabloid reporter who specialized in entirely fictional "top-of-the-head" stories. He claimed it was almost impossible to produce a story which did not receive some believing response from some readers. He described, for example, a story he had written about a 67-year-old woman who married a 17-year-old boy, after being rejuvenated by using a Guatemalan jungle plant with magical "fountain-of-youth" properties (Hogshire 1989). He explained how he put the story together from his editor's headline, using a convincing journalistic structure and a vivid imagination:

> In the story, I tried to make it as mundane as I could—you should do that with all top of the head stories—make the conversation as mundane, the people's lives as mundane as you can. That's how you can sound most realistic. The name of the god who supposedly gave the plant was an anagram for the Spanish word for peanut—I made him a toucan-headed god, he's the one who brought the plant according to Indian lore. And then I didn't exaggerate too much, I said old people were mentioned in Cortes's diary, they lived more than 120 years or something...and then I located the university in the middle of this very dense, totally uncivilized jungle, and we used a couple of fake photos of vaguely Hispanic-looking people. (Hogshire 1992)

The response was immediate, and apparently typical, as Hogshire received dozens of calls and letters begging to know the specific location of the plant. People argued that "they needed to find this plant because they would know how to market it, whereas the Third Worlders couldn't be trusted to do this." Most extreme was a man setting himself up in business as a herb dealer: "He would call up and cajole, call up and promise to give me other stories, he hinted there was a lot of money to be made on this thing... he called up pretending to be somebody else, he wouldn't quit..." Finally the man went down to Guatemala, and ended up calling Hogshire from Guatemala City begging for directions to the miraculous plant.

Clearly not all tabloid readers are that credulous; readers take what they want from the papers, and this varies greatly. I did conclude that "resistance" is an important element in their reading, in the form of a resistance to dominant values—an awareness, for example, that they "should" be reading about news and current affairs, but find these boring and irrelevant. Tabloids glory in stories about the bafflement and discomfiture of politicians, establishment science, or liberals who control education and the media, and readers found these tales among the most satisfying. Yet even the kind of stories Glynn (2000) and Fiske (1992) regard as clear parodies (and thus reflective of audience "resistance"), can be and are taken seriously by readers who want to believe them. Tabloid writers often try to outdo themselves and their colleagues in producing outlandish, "top-of-the-head" stories that will still receive a serious response from some readers. This kind of cynicism hardly suggests a medium whose goal is to help readers resist the establishment. In fact, it acknowledges how far most tabloid readers are from the center of real power, and seems closer to Schroeder's (1982) characterization of tabloids as "a fetish of hope for the helpless" (p. 180).

Furthermore, even those audiences who are the most clearly active, playful, and creative are necessarily constrained by the political economy of the media industries. Fiske (1989a, 1989b) offered the necessary antidote to the notion that the "culture industry" determines what "we" will see, hear, and read, turning us into mindless dopes. He sensibly argued that something only becomes popular if the people make it so, and that of all the thousands of TV shows, movies, and so on contemplated or made, only some become truly popular. And these are not always the ones the industry expects to be successful. That is true, but of course it is not the whole story, because it does not take into account the marketing forces that create an enormously lopsided balance of power. Are U.S. soap operas successful around the world because they are instantly appealing in all cultures, as local audiences busily reinterpret them within their own contexts? Maybe, up to a point. But we all know that the central reason they are shown worldwide is that they can be bought much more cheaply than local programming can be made. Viewers "choose" them, but often it is a Hobson's Choice.

Even dedicated fans, devoted to pleasurable consumption, work within constraints. The *Dr. Quinn* fans with whom I worked were among the most creative one could find. Yet when the network executives decided to "tweak" the show to improve the male demographic, they had to deal with the intrusion of the hated "Daniel" into their show, and the perceived "dumbing down" of the female lead. Fan fictions can tell their own stories, but the reality on their screens is hard to counter. And when the young male demographic became the Holy Grail for the network, the DQ fans were doomed. They organized, they collected money, and they took their case to the press, but they never got their show back. They learned the truth about the political economy of network TV, that "not everybody is wanted by advertisers, so a ratings firm must try to exclude those persons who are not in demand" (Meehan 1990, 127). As Meehan goes on to explain, "Thus it is macroeconomic structure—not taste, not training, not temperament—that determines who counts in television" (p. 132).

As the Internet has risen to prominence, much has been made of its newly-liberating power to allow audiences to make up their own minds about media offerings, and perhaps to influence the power structure. Internet "buzz" was credited for the success of such unlikely vehicles as the *Blair Witch Project,* and it certainly seems likely that the organized online fans had a role in ensuring that two DQ movies were made after the cancellation of the series. The Internet has undoubtedly opened up whole new worlds of pleasurable consumption for those fans who have easy access. At the same time, the media industries have been quick to commodify this new space too—setting up manufactured "fan" sites to create buzz, and offering opportunities to buy products even before a movie is released.

THE WORLD AS FUN HOUSE: CONSUMERS OR CITIZENS?

The Internet has been the site of now-famous battles over music file-sharing and the use of copyrighted images and clips from movies and TV, and these battles have often been characterized as power struggles between the oppressed people and those with media clout. Indeed at one level they are. The increasing consolidation of the media industries is truly frightening; as McChesney (2000) puts it, "The system ... does not exist as a result of popular will ... The media system exists as it does because powerful interests have constructed it so that citizens will not be involved in the key policy decisions that have shaped it" (p. 15). So perhaps any subversion of that system must seem welcome. One DQ fan, having read my paper about the list, raised questions about my reluctance to characterize fan resistance to the cancellation as a fight against oppression. She agreed that it was a stretch to define their actions as in some sense political, while pointing out, as others did, that they now saw their movement as being "feminist" in a genuine sense. "Mary" had quickly understood that the way to reach the network was to reach the program's sponsors, and enlist their support. She wrote:

> This list is the only reason I entered the campaign to save DQMW, and tackled the world of sponsors. And while not getting the TV shows that appeal to you because you are "too female, too old, too rural" may not be the greatest oppressive ideology we experience in life, it does seem to me to be symptomatic of who is running things and the dangerous precedent they are setting for future programming for child viewers. And that can be argued to lead to the oppression of "violence" that we all feel invading our lives.

I believe she is right, in that if we accept the old feminist adage that "the personal is political," we need to be alert to how personal oppression reflects and represents systematic social and political oppression. If fighting for a TV show one values leads in turn to a sharpened sense of feminist identity, this might spill over into other aspects of life. Indeed, other DQ fans have talked about how their new knowledge of Native American history inspired them to become active in Native American causes; a current list project is a fund-raising activity for a Native American foundation.

Nevertheless, this kind of transformation from consumer to citizen seems rare. The founders of file-sharing sites were hailed as revolutionary heroes, using "subversive" technology, when in reality, they were merely offering more opportunities for consumption for those with the resources to benefit. And Napster's founder now has a deal with a media conglomerate. Resistance means fighting for the right to have fun.

Nevertheless, perhaps having fun can also be part of being informed and participating in our society—which is where we return to the cultural role of news and information, and specifically its future. Many young people (at least those with the resources to do so) use the media in creative, active, and very pleasurable ways. As Drotner (2000) points out, young people's media use is very intertextual–they are not "the TV generation" or the "computer generation," but simply the "media generation," who are surrounded by images, and images of images at all times. They are comfortable with images, and find them pleasurable. And they often seem relatively unconcerned about the relationship of image with "reality." At an anecdotal level, I have been interested in (and admittedly worried about) a noticeable change over the years, as I teach a graduate media in culture class. At one point in the semester, we discuss the issue of digital manipulation of images, and its implications for journalism. Even 10 years ago, classes were generally united in their opinion that this is a troubling trend that raises important ethical issues. Older graduate students still think that way. Yet in the last few years, many younger students have taken a different view, which could be summed up as "It's no big deal. We know all images are altered and manipulated, so it's up to viewers to reach their own conclusions about whether something is 'real.'" For some, the conceptual chain that links image to reality is stretched or even already broken.

Now maybe this doesn't matter when we are talking about pure pleasure. Playing with style, fashion, music—the continuous, refashioning of images that active audience researchers have found–is fun and undoubtedly creative. In consuming media, people are able to think through issues, refashion themselves, and construct "imaginative strategies" for their lives (Fisherkeller 1997). However, as more and more critics are beginning to ask (see most recently Crenson and Ginsberg 2002), should life be only about consumption and pleasure? Is it not also important that people function as citizens, and not merely as consumers? In our ever-more-pleasurable exploration of exciting media images and messages, are we simply sitting back and allowing those in power to make our decisions for us, whether these decisions are about "homeland security" or the necessity of war? As Hallin and Gitlin (1994) suggest, when the Gulf War became a tightly controlled media spectacle, it also became a pleasurable, almost aesthetically-pleasing entertainment that encouraged audience excitement, patriotism, and vicarious participation. Its stories about individual soldiers doing their part for freedom touched many viewers, spurring them to write letters, send gifts, weep, and celebrate. The "war against terror" in Afghanistan was also a mediated, "feel-good" war, as we watched more smart bombs, sophisticated technology, and triumphant soldiers, all described with breathless excitement by "personality" journalists like Geraldo Rivera. There can be no doubt in anyone's mind that media coverage of "Enduring Freedom" was highly successful in uniting much of the population, and that for many Americans, it was a positive, pleasurable experience (see Zelizer and Allan 2002).

In a similar vein, I explored in chapter 2 the pleasurable and culturally-useful role of scandalous news in everyday life, arguing the importance of understanding news as storytelling. The great strength of news in this role is its ability to help us connect stories of personal lives and emotions to our own concerns and those of the wider culture. But of course the weakness of the personal, entertaining tale, whether about scandal or about war, is that it can become just another story that engages our emotions, while failing to help us understand the complexity of issues that ultimately affect our lives. And when stories about the "real world" become effectively interchangeable with stories from fiction, perhaps we are in danger of losing touch with the functions of citizenship.

Let me illustrate this point with some further discussion of my study of audience understandings of news that contributed to the analysis of scandal discussed in chapter 2. Narrative appeal is at the core of the pleasure of the personal story, whether that story is a scandal or anything else. Also important (and not of course unrelated) is the vivid verbal or visual image. Television news has always had the advantage of the visual image, and has long been criticized for sensationalizing it, as was print journalism before it. The debate over appropriate images is wide-ranging, but one often-cited symptom of the decline in serious news, apart from the rise of the "tabloidized" story (Sparks and Tulloch 2000) is the way the image has crowded out rational analysis. Indeed in television, the existence of a striking image will actually determine whether a story is used or not, especially on that ratings-driven genre, local news, which is watched by far more Americans than national news. American local news is a litany of unconnected, often violent images, frequently of events that have no immediate relevance to the region in which they're shown—freeway pile-ups, fires, police chases, and stand-offs. A 1995 survey by Rocky Mountain Media Watch of 100 local television broadcasts found that 42 percent of airtime was devoted to such "mayhem" (Krajicek 1998, 27). In addition, local news anchors are encouraged to display visual aids whenever possible—not just the traditional charts and graphs, but such items as cans of beans to illustrate a consumer story (Heider 1998, B8). Going one step further, "tabloid" TV shows from their beginning were not even constrained by the availability of actual news footage. Instead, they developed the re-enactment or dramatization, in which events are recreated by actors for the cameras. The re-enactment was probably first used in August 1986 when *A Current Affair* aired a dramatization of events leading up to the "Preppy Murder," a widely-publicized case in which a wealthy young man was accused of killing student Jennifer Levin (Krajicek 1998, 42). In the following decade, the technique became a stock feature of reality shows and syndicated news magazines, and is widely cited as one of the key elements in the oft-lamented "tabloidization" of news (Langer 1998, Sparks and Tulloch 2000). For a while, it seemed that the re-enactment was one feature that separated tabloid and serious news. But recently, dramatizations have started to creep into network news magazines and local news, particularly in crime re-enactments like *America's Most Wanted,* the popular *Forensic Files,* and the now-ubiquitous news coverage on channels like

Fox. Gradually the dramatization is beginning to be seen as just another way to enhance the story, to bring it to life, and essentially to increase the pleasure of the audience.

In my research, I was interested in how people perceive such techniques. While it would be unsafe to draw major conclusions from a small project, I was intrigued by a generational difference that emerged. Middle-aged and older people did not like re-enactments, seeing them as somehow cheating, as detracting from the reality of the news. Typical was a woman in her forties who appreciates the dramatic value of the technique, while clearly distrusting the effects:

> Well, it's a great tactic, you know, because there's nothing that sticks in a person's mind longer than the picture of something...But you know my thought is...how do they really know?...I mean, were these people unkempt and unshaven, or were they really pretty and sexy-looking, and smelled like Aramis? I don't know...if I'm just looking at it and not thinking about it, it would make it more real because of the visual effects.

This woman points to the fact that the dramatization can actually seem "more real" (or "hyper-real," as Boudrillard would put it), unless the viewer actively works against seeing it that way. Younger people, however, seemed less bothered by the technique—in fact it seemed to enhance their appreciation. Thus a male college student comments:

> It's sometimes hard to visualize how things happen, and when they do re-cap 'em, it does help . . . For example, I was watchin' (*Rescue*) *911* one time, and an ambulance was following a car . . . the mother had a seizure and she fell asleep at the wheel, and there was a baby in the front seat, and they totally recapped this, the story, and the thing was, she was like bouncin' off curbs, and swervin', and one of the ambulance drivers got out and tried to run and catch the car . . . I mean, you're like, wow, that really can happen. And I'd say somethin' like that was really beneficial to understand the story. Kind of like a movie.

Later in the interview, this young man was talking about how he now believed the government is hiding things from the people. When did this come to him? "Oh, yeah, after I saw *JFK*. That really changed my view." We see how something that is actually a recreation, told for dramatic purposes, is perceived as more convincing than either traditional news or history. Another male college student agreed. "I like it because it lets you know exactly what's going on . . . whereas words you can still, you know, imagine things." When asked if this might create a danger that the audience might not be seeing what actually happened, he replied, "No ... They have no reason to really lie."

If there really is a generational difference, is it because younger people are indeed the media generation? Are they more easily persuaded by the truth of visual images, which are seen as somehow transparent, obvious, unmediated? Perhaps the "realness" of something may be more tied up with the impact of the images, rather than whether those images correspond to any outside reality. If that is true, it is a worrying thought, particularly as we more deeper into an era in which technology makes image manipulation more and more sophisticated—and constructed images and documentary images become seamlessly intertwined. For at the same time, the younger people in my study were also the most likely to be cynical and untrusting of *any* news or any "facts" generally, reflecting a kind of relativism that is often seen as the hallmark of the postmodern age, and which I have also observed in my own students. From this perspective, any viewpoint, any "fact" is as likely to be as "true" as any other. Perhaps the fact that is presented with most visual impact, the one that can "break through the clutter" as advertisers say, is the one that wins out.

At the same time, "reality" does seem to be an important value for viewers, young people included. If something is real, and happening to actual people, it carries a ring of authenticity. The enormous popularity of shows like *Cops, America's Funniest Home Videos, Survivor, Fear Factor,* and the endless video compilations aired by Murdoch's Fox network, such as *When Animals Attack,* attests to this. Producers can get away with poor production values if the effect is "real." Yet at the same time, young viewers seem happy to accept the ability of producers to stage or re-enact events to make them more real than real. Like the young man quoted above, a young woman in the study argued that "Everything that was said (on the re-enactment) was things that really came out of one of the people's mouth at one time during the case ... I mean, like ... they can't twist it too much because they have to give people rights."

Grindstaff (1997) describes how talk shows respond to the very audience demands I have been discussing: "Producers want 'real' people to provide first-hand testimony ... the more emoting the better. They do not want distanced analysis, or complicated discussions of politics or law" (p. 192). Yet at the same time these real people can be hard to handle: "These are the guests who, in some ways, pose the greatest challenge to producers because the very qualities that make them 'real' make them more difficult to manage ... The pressure of deadlines, the nature of the topics, and the performances required of guests actually push producers toward people who *are* media savvy, have had prior talk-show experience, and may even be actors faking their stories" (p. 189).

We know that news audiences generally prefer lively, dramatic, human interest stories over news about political and economic issues. As I have argued, this is not necessarily a terrible thing. News can and should be pleasurable; conversations that viewers and readers have about news stories serve to bind people together, and give them common topics of conversation in a world in which common ties are getting fewer and fewer. News stories of scandals, even of such trivial tales as Amy Fisher and Joey Buttafuoco, offer an entry point to everyday discussions of morality, boundaries, and appropriate behavior. And although it is often argued that thinking about trivia prevents people from learning about important issues, there is little evidence to suggest that if people were not discussing Amy and Joey, they would necessarily be discussing international politics.

Audiences do make active choices about the news they can use, and berating them about what they should be interested in is unlikely to have much effect. However, audiences can only run with the stories they are given; as Sparks (1992) comments, "the sense which people can make of newspapers depends at least in part on what the journalists have actually written in the first place" (p. 37). The current climate of tabloidization in news is a product of the dynamic between audiences, journalists, and economic forces. Stories like Amy Fisher or Monica Lewinsky are demonstrably popular, and earn ratings. But it is vital to remember that they are also cheap and easy for news outlets all over the country to run—they are delivered on the wires, and they can fill hours and inches of news space profitably, for very little reporting effort. Similarly, war stories from designated "news pools" require almost no journalistic effort, yet are freely available for saturation coverage. A newspaper or broadcast that is full of scandal and war does not have a staff digging to uncover financial mismanagement, corruption, or even relatively routine but significant local news. In that respect, the pre-packaged human interest story, especially the big, national story, *is* pushing out diversity of information from the news media. And if stories are not placed on the agenda by the media, the audience cannot discuss them.

The danger as I see it is not the personalization of news in itself. Tomlinson (1997) is optimistic about the power of the personal story to have impact and genuinely inform: "Those experiences that penetrate deepest into our lifeworld are the ones that can be imaginatively incorporated into this on-going narrative of self-identity" (1987, 73). He argues that journalists must try harder to interpret important national and global issues in personal terms in order to engage audiences. Indeed, that does happen. We may recall a famous photograph of two lovers in the former Yugoslavia, who died under gunfire while trying to reach safety. It was certainly emotional and sensational, yet that photo and the explanatory story introduced many Americans to a reality of which they had only been vaguely aware.

But even this kind of fairly superficial reporting requires resources and the will to try. It is cheaper and easier to buy wire stories and pictures that do not require interpretation or analysis, or to focus on shocking or heart-warming tales generated locally. The real danger in the transformation of news is that the trend toward personalization may become the *only* way to tell a story, and that these stories become increasingly disconnected from a larger context. As Sparks points out, "The simple reality is that the nature of the social totality is neither constituted through immediate individual experience nor entirely comprehensible in its terms . . ." (1992, 41). American television is an intensely competitive environment, in which multiple channels must be constantly supplied with material. It takes money and effort to contextualize and analyze stories, and it does not necessarily pay off in greater audience interest. More and more, the U.S. electronic media offer stories whose aim is simply to engage our emotions, where indeed "the 'personal' obliterates the 'political' as a factor for human behavior" (Sparks, 40). Nowhere is that clearer than in the shrinking and increasingly trivial coverage of international news, particularly on television, as discussed by Moeller (1998), and mirrored in the United Kingdom in a recent report produced by the Third World and Environment Broadcasting Project (2000).

And in a climate that validates personal experience and pleasure over logic and reasoned argument, it becomes equally valid to pay attention to any and all personal views, no matter how uninformed, bigoted, or irrational. So in the United States we have seen a proliferation of shows in which people share their experiences with fatalities and fetishes, apparitions and angels, all without context or comment. The *Jerry Springer Show*, in which foul-mouthed guests come to blows under the auspices of themes like "I Slept with Three Brothers," or "Men who Dress Like Babies" became for a significant period the top-rated daytime talk show, proving immensely popular with young people, who egg the guests on with chants of "Jerr-y, Jerr-y," while clearly enjoying themselves immensely. The show only reached the boundaries of acceptability when producers canceled a program on guests who practice bestiality, titled, "I Married a Horse" (Deggans 1998, B2).

Springer's core audience returns us to our starting point with young people. Younger audiences are especially unlikely to pay attention to news. A 1996 Gallup Poll showed that people under 50 were far more able to identify sports and entertainment figures than politicians and international figures (for example, 15 percent could name the prime minister of Israel, while 75 percent could name the host of the

Tonight Show). Older Americans were still relatively ignorant (only 70 percent could name the Vice-President), but their recognition of popular and hard news figures was much more even (Newport 1996). The Pew Research Center reported in 1996 that news viewership had declined among people under 30 more than among any other group, with only around 20 percent watching either network or cable news.

Some of this may simply be a difference in life-stage; older people may have more time to pay attention to the news. But more likely it is both a reflection and a cause of the changes in the news business. The contemporary United States has seen an astonishing proliferation of news outlets—talk shows, TV magazines, TV "tabloids," cable networks, and most recently, the Internet. All are fighting for audiences and competing to meet the proven demand for spectacle. Many of the younger generation, with access to an incredible array of entertainment, appear to have given up even the pretense that "being informed" is useful and necessary.

And U.S. journalism itself sometimes seems to be giving up the effort to do anything but tell stories and provide spectacle. Respectable newspapers plunge headlong into the competition, demonstrated in the saturation coverage of the death of Princess Diana, and the Clinton-Lewinsky scandal. Indeed, in the early days of the scandal, spurred on by rumors emanating from the Internet and elsewhere, the press published information without attribution, much of it later proved wrong. Although some media then stepped back and issued a *mea culpa* (see, for example, Dahl 1998), in some respects the damage was done. For the audience relationship with the press is a complicated and ambivalent one. Although the taste for sensational news is clear, at least for now the public is still torn between that taste and a perception that the press should be more than amusing storytellers. The Pew Research Center reported in 1998 that public criticism of the press for inaccuracy, unfairness, intrusiveness, and sensationalism is at an all-time high. The irony is that the more the media try to respond to audience demands, the more they erode their increasingly tenuous position as a focus for serious public debate. But of course in an increasingly consumer-based culture, if the "new news" reaches audiences and makes money, it has demonstrably succeeded.

The argument has been made that the explosion of news outlets has democratized news, making it more responsive to what people want. Indeed, in less media-saturated cultures, the emergence of market forces in news, in such places as Mexico (Hallin 2000) and Hungary (Gulyas 2000), undoubtedly has had a liberating effect, offering more choice and more diverse viewpoints. In the United States, the deluge of new information may be little more than an illusion, at least when we consider the traditional media outlets of radio, TV, and the press. It appears to offer variety, but it is the same old stories going round and round all the time. In theory, greater popularization leads to more democracy, but inherent in that theory is the notion that as more people's voices are heard, they will gain access to the wider political process. That hope may be vain; Grindstaff (2002) suggests, for instance, that while talk shows love to use "ordinary people," who are all too eager to air their grievances on national television, the format is guaranteed to encourage emotionalism and volatility, actually further marginalizing them as entertaining but freakish "trash," who are literally discarded after their brief moment in the spotlight. As I discuss above, blanket war coverage produces the same "more voices, less information" effect.

BUT WE CAN TALK BACK NOW

So how do we as audiences break through the grip of the media conglomerates, talking back to media as active citizens, rather than mere consumers? The conventional media seem to offer relatively little opportunity, and once again attention turns to new technology, and specifically the Internet. In discussions of media reception, it seems we constantly find ourselves moving between good/bad judgments: "On the one hand, X seems to offer opportunities for audience creativity, but on the other hand, Y places significant constraints." The Internet is no exception. The explosive growth in access to all kinds of information has the potential to interrogate the power of the media conglomerates as nothing before has, and clearly many are taking advantage. At the same time, we also see how quickly the Internet, like the rest of our culture, has been colonized by corporate interests, and has become very much a commercial space: "One consequence is that civic discourse has given way to the language of commercialism, privatization, and deregulation. In addition, individual and social agency are defined largely through market-driven notions of individualism, competition, and consumption. As such, the individual choices we make as consumers become increasingly difficult to differentiate from the collective choices we make as citizens" (Giroux 2002).

Clearly this is not the place to enter in any depth the now huge debate among media scholars about the nature of the Internet, and what (if any) impact it is having on contemporary culture. Many aspects of that debate, as I discuss in chapter 3, are as polarized as long-standing disagreements about "media effects." But I believe it is worth discussing, if briefly, the role of electronic communication in offering possibilities for active civic participation beyond the role of the consumer. According to the Harvard Business School's website, Business and the Internet: Strategy, Law and Policy (1998), the Internet has become a major source of news; of 20 million users of the Internet, 53 percent are news consumers, and MSNBC, the online arm of NBC news, has over 4 million visitors per month. Furthermore, the growth of interactive chat rooms, news discussion lists, bulletin boards and so on, has opened up the possibility for audience participation in the development of the "story" on an unprecedented level. In theory, anyone can be heard, with any opinion (even if one clear consequence is that we can now talk endlessly about Bill and Monica with even more people, as a quick look at the favored topics on news boards will confirm). And we are no longer restricted to news sources that are easily available; as an expatriate today, I have an entirely new ability to keep up with British current events, or to find different international spins on U.S.-defined stories, without moving from my desk.

The significance of this should not be belittled. The potentially subversive power of the Internet leads to attempted crackdowns by repressive governments, yet people around the world still find ways to find out what is happening, and connect with like-minded people. In this country, the web-log or "blog" has become a fascinating way for individuals to document their daily thoughts and responses to current events—even for media scholars to take their critical analyses to a wider audience, as Kellner does in his blog, "Critical Interventions" (www.gseis.ucla.edu/courses/ed253a/blogger.php). Blogs have been hailed as revolutionary: "Blogging is changing the media world and could, I think, foment a revolution in how journalism functions in our culture" (Sullivan 2002). Sullivan goes on to argue that blogs "seize the means of production. It's hard to underestimate what a huge deal this is. For as long as journalism has existed, writers of whatever kind have had one route to readers: They needed an editor and a publisher. Even in the most benign scenario, this process subtly distorts journalism. You find yourself almost unconsciously writing to please a handful of people . . . Blogging simply bypasses this ancient ritual." Blogs have an exhilarating capacity to deconstruct the texts of the powerful. For instance on his "Way Nu" blog, Jonathan Peterson (2002) offers a detailed analysis of a speech made by Fox CEO Peter Chernin, in which Chernin presents a cheerful analysis of the wonderful media future, suggesting, for example, that media are becoming more diverse than ever. Peterson juxtaposes Chernin's statements with his comments: "The media industries certainly aren't diverse and they are getting more consolidated all the time," linking this directly to sites with information about media monopolies. When Chernin blames technological piracy for the supposed dire problems of the movie industry, Peterson again offers detailed source information, while countering:

> U.S. box office hit $8.4 billion in 2001, a 9.8% increase over the previous year. Worldwide box office receipts for feature motion pictures have grown from $1.2 billion in 1970 and $2.8 billion in 1980 to over 15 billion in 2001. This increase is all the more remarkable because ancillary markets such as home video, cable and (foreign) television markets have undergone explosive growth during this same period. Hardly an industry in crisis is it?

Blogs can, and do, provide and link information that equips audiences with knowledge to question media conglomerates and other powerful interests. Of course they are plagued with the problems that characterize the Internet as an entire system: How can we trust information? Above, I offered figures from the Harvard Business School, which suggests trustworthiness. I did not mention that they came from a web site created by students in a class, a point that could potentially tarnish their credibility. For Sullivan, one of the glories of blogs is that they offer opinion un-framed by the bogus credibility of "official" news organs: "Readers increasingly doubt the authority of *The Washington Post* or *National Review,* despite their grand-sounding titles and large staffs. They know that behind the curtain are fallible writers and editors who are no more inherently trustworthy than a lone blogger who has earned a reader's respect." What this points to is the enormous onus the new information deluge places on the audience; we now have to work to assess the context and credibility of everything we read. Or perhaps exhausted, we can subside into the blanket relativism I have worried about in my students; if nothing can inherently be trusted, then anything is as likely to be true as anything else. And how does that translate into action?

Indeed, blogging is also seen as one way to control that information deluge. Internet news enthusiasts point to the way many news fans already tailor their news by subscribing to sites like MSNBC and requesting "custom" pages. Kling (2002) offers "a model of blogs in which blogging serves as a filtering mechanism in the dissemination of information." This means we pay attention only to blogs that address our concerns: "If the filtering system works well, I get to read lots of economic insights, and I never have to read anything about, say, Olympic figure skating." This process is not fundamentally different from the way people already consume news, in that we filter out the stories that bore us, as both my research and that of others (e.g., Graber 1984) indicates. One of my conclusions in the tabloid research was that tabloids do not seem to instill beliefs or opinions in their readers, but rather they reinforce those already held. In Britain, most people who read the *Guardian* do so not because they are Tories trying to challenge themselves, but because they are liberals who want to confirm their opinions. However, in a media market that has already become highly specialized and segmented, the idea of personalized electronic news has an ominous ring to it. It's the same story yet again: On the one hand, the Internet clearly offers unprecedented opportunity to talk back, and to join with others in resistance. On the other, it also allows us to retreat into an individualized world inhabited only by people who think exactly as we do. The choice is up to us; do we engage with others in the ever-expanding public square, or do we simply use it as another opportunity to buy toys for the fun house?

THE FUTURE OF MEDIA ETHNOGRAPHY

Livingstone (1998) recently wrote that audience research is "at a crossroads," and she suggests scholars need to step back for a moment and assess where we have come, and where we need to go. She argues that we must resist a "canonical" approach to audience scholarship, and move toward conceptualizing audience study as part of a larger project of understanding mediated culture, establishing "a research agenda . . . that connects audience research with production/texts/context research as firmly as actual audiences are, inevitably, connected with actual production/texts/contexts" (p. 196).

This book is my attempt to contribute to that agenda, although I would like to broaden its conceptualization a little. I believe the move "beyond audiences" is not something unique to media studies. Rather, we can see it as part of the more general agenda in anthropology to move "beyond cultures." Of course anthropology still holds the concept of culture as its core. But just as media scholars recognize that "the audience" is not a discrete, bounded entity that sits still to be studied, anthropologists increasingly recognize that the bounded "cultures" we used to study are conceptually dissolving. Among others, Appadurai (1990) sets the agenda for a study of global cultural processes and connections rather than "cultures" as such, as for instance we trace the reach of international "mediascapes." As Marcus (1998) points out, Appadurai does not offer suggestions as to how we might study "scapes" ethnographically. Indeed, a central struggle in anthropology mirrors the conceptual problems in media audience studies: Where does ethnographic study of the local belong in broad explorations that acknowledge the interaction of the global, the national, and the regional, and where the influence of corporate power is felt at the level of the individual?

Marcus (1998), one of the defining figures in anthropology's "crisis of representation," argues that we must not abandon ethnography, and we certainly should not become embroiled in the "more-ethnographic-than-thou" debates that enshrine long-term participant-observation as the *sine qua non* of "real ethnography." Instead, he advocates the same kind of opening-up of methodology for which I argue in chapter 1. And at the same time, he asserts that we must learn how to do ethnography in more dispersed, "multi-sited" ways that take into account the local experience, while also understanding the way that experience is constrained by larger forces. Such new ethnographies "literally move over discontinuous realms of social space in order to describe and interpret cultural formations that can only be understood this way. Now they must understand the operations of institutions (e.g., information systems, corporate cultures, media technologies) as much as the modalities of everyday life lived in communities and domestic spaces, which have been the most usual sites of anthropological study" (1998, 240). Thus we find anthropologists experimenting with different, and certainly multi-sited ways to evoke experience in a globalized setting, as Edwards (1994) does in his study of Afghanistan. As Edwards writes, "Despite our best efforts to be truthful, traditional modes of organization and articulation can and often do lend to what we construe as 'our research objects' a formal logic and coherence that bears little or no relationship to our original experience of those objects" (p. 358). He tries to deal with this problem by presenting his research in deliberately disconnected ways, combining traditional "field" accounts with discussion of mediated narratives, Internet communication among displaced Afghanis, consideration of the transforming effects of personal technology like cell phones and video, and an attempt to put actual events in Afghanistan in their global context. The result is far from a traditional ethnography, yet it retains the essential defining features of an ethnographic "way of seeing."

Clearly, this multi-faceted conceptualization fits very well with the notion of understanding media audiences as simultaneously creative and constrained, and never being in one place. Guided by this perspective, we need to continue studying real people and their interactions with the media, but let us attempt to understand how their daily choices are limited not only by their own social and economic circumstances but also by the power of inscription held by media producers. As Grossberg (1997) argues, the study of media audiences in this country "must confront not only the very real and often dire political conditions of the United States, but also the important place of the media in the current reconstruction of those conditions . . . We need, then, not a theory of audiences, but a theory of the organization and possibilities of agency at specific sites in everyday life" (p. 341).

Marcus (1998) offers several ways to approach the multi-sited ethnography, any of which could and sometimes have been useful in addressing the role of media in everyday life. For instance, he points to studies that "follow the people"–for instance, in ethnographies that, while perhaps focusing on particular ethnic groups in particular places, draw out the wide connections between their local experience and the wider diaspora in which their cultural identity is located. In media studies, we might cite those by Gillespie (1995) on the media use of Indian people in London, or Miller and Slater's (2000) study of Trinidadian identity on the Internet (although Marcus himself does not mention these). Next, Marcus suggests "follow the thing," citing Mintz's (1985) defining study of the cultural history of sugar, as well as studies of music, cars, art, and dance, which focus on particular forms, mapping connections from the form across national or global culture.

Marcus also suggests we might "follow the metaphor," as Martin (1994) does in her complex study of ways of understanding the body's immune system across time and space in the United States; or "follow the conflict," as Ginsburg (1989) did in her study of how abortion was contested in a small American community; or "follow the life," by focusing on individual life histories, again tying them to larger forces. Perhaps significantly, none of the studies cited by Marcus are identified as "media studies," although almost all of them acknowledge the powerful role of the media in constructing the cultural reality of any of these themes. However, his recommendations explicitly inspired the continuing project of Abu-Lughod (2000), whose work on television in Egypt is defined as a multi-site media ethnography. Abu-Lughod reflects on how different it is to study the dispersed "audience" for TV:

> What kind of fieldwork is needed to track media in lives? I find it intriguing, for example, to compare the fieldnotes and material I have for my television research with the small dog-eared notebooks and simple audio cassettes that resulted from my more localized research among the . . . Bedouin. Now I have different sized notebooks that are filled with notes taken in very different places . . . observations and conversations with people are recorded alongside summaries of plots and bits of dialogue from . . . soap operas . . . I carry back from Egypt video cassettes of television programs and piles of clippings from newspapers and magazines, some with movie stars on the covers. (p. 26)

Abu-Lughod goes on to assert the crucial importance of then exploring the power dimensions of media, concluding that an ethnography of Egyptian television tells us "what it means to be a nation at the cross-roads of Arab socialism and transnational capitalism with an intelligentsia promoting modernist and developmental programs and ideals through a state-controlled medium to a varied population, large segments of whom remain uneducated and marginal" (p. 26). This kind of contextualized, nuanced reading of the role of media, moving from "audience" to producer, and the forces working on both, offers a challenging model for ethnographers as we move "beyond audiences."

Finally, Marcus argues that there is still a place for detailed ethnographies of local experience. Such ethnography "may not move around literally but may nonetheless embed itself in a multi-sited context" (1998, 95). It is here that more focused studies of media interaction fit, assuming they attempt to move beyond the immediate moments of reception. This is where I try to situate my own work, as exemplified in the studies in this book, and in my work on supermarket tabloids. In that project, while I discuss (and arguably celebrate) the activity of the readers, I also attempt to show how the political economy of the tabloid industry, as well as the ideology of those who produce the papers, ensures particular ways of seeing and certainly constrains the creative options for consumption.

CONCLUSION: REAL PEOPLE IN A MEDIA WORLD

We are indeed at an important moment in ethnographic studies of the media. Finally, after traveling rather different paths, and still influenced by different research traditions, anthropologists and other media scholars are coming together with a common concern–to understand the immensely complex ways in which we live in a media world, whether our immediate locality is a city in Africa or small town U.S.A. This is not to suggest that the media world is the *same* in different cultural milieus. Indeed, central to the challenge of audience research in the West is the development of an adequate understanding of the uniquely media-saturated cultural conditions in an advanced capitalist society, as opposed to the equally unique conditions described by, for example, Abu-Lughod in Egypt (2000)—while at the same time, each is also impacted by global economic and political forces. In any context, methodological innovation and flexibility are now keys to continuing progress in this interdisciplinary project.

Team projects, long relatively rare in media ethnography, are beginning to emerge as important possibilities for the study of contemporary U.S. life, as American anthropology increasingly turns its gaze to home (di Leonardo 2000). For instance, ethnographic work on the media in family life is the focus of two innovative foundation-funded centers, associated with two disciplines that are still largely unconnected—the University of Colorado's Resource Center for Media, Religion and Culture (affiliated with the School of Journalism and Mass Communication) and the University of Michigan's Center for the Ethnography of Everyday Life (affiliated with the Department of Anthropology). Both centers are producing rich, ethnographic accounts of living mediated lives in contemporary America, and both approach their task in a very similar way. Team-based, broadly ethnographic projects like this, which do not focus on particular moments of reception, perhaps hold the most potential for a richer understanding of not only the active "Kathys" of our culture, but also the more passive "Kevins" and "addicted" young people.

At the same time, narrower text-focused studies still can offer insight about particular audience moments. For instance, anthropologists rather belatedly (and disconcertingly) joined the "multiple decodings" movement when they discovered that students who watched ethnographic films in anthropology classes often emerged feeling more rather than less negative toward the "primitive" cultures they depicted (Martinez 1992, 1996). I find it fascinating that, as researchers so many of us understand that media messages and images are not received in predictable ways, but as teachers we often use film and video in classrooms and unproblematically assume our students uniformly understand them the way we do. Studies of media use as a teaching tool, interrogating the interaction of teacher, text, and student in a specific classroom moment, would add an intriguing dimension to the canon of media reception scholarship. No doubt there are many other examples of studies that could begin from a narrower perspective than "everyday life"; my purpose here is simply to argue for a genuine sense of openness and innovation in both subject matter and methodology.

I believe the project of understanding media in everyday life is not merely "academic," in the sense the word is used to connote insights that are of interest only to a few specialists. An understanding of media's potential to penetrate virtually all aspects of culture is crucial—to educators, to policymakers, to scholars in all disciplines, and to parents. Popular discourse on the media is oddly bipolar; almost everyone enjoys, consumes, and talks media, and almost everyone seems to fear and condemn media hegemony. Rich, ethnographic media studies can help inform that popular discourse; demonstrations of audience activity can make us feel less helpless and more powerful, and can perhaps help us find ways of, as Livingstone (1998) puts it, "bringing such micro-level studies under the umbrella of citizenship" (p. 199). Meanwhile, analyses of media passivity and dependence may also serve to shock us out of complacency, and help us find ways to understand and control our dependency. Dardenne (1994) offers an often bleak and depressing account of using the "media deprivation" exercise, noting that many participating students genuinely "think they are addicted to mass media...without media there is nothing. Many use media extensively to eliminate silence, boredom, and even thinking" (p. 72). Ending the exercise is a relief, as students "noted how happy they were to be back in touch with reality, as though mass media were reality and their own lives something else. In fact, many argued that medialess lives are not real, certainly not normal" (p. 73). And indeed, that's probably true in today's Western society, in which "mass media provide companionship, facilitate social intercourse, and make other activities bearable" (Dardenne 1994, 72). Anthropologists and other non-media specialists still draw rigid distinctions between culture as experienced through interpersonal communication and through media. Yet the two are inextricably mixed: "At home, at school, *and* through television, each student learns about the interrelated power and problems of gender, race/ethnicity, class, and other identities. Likewise, they learn about standards for success and for group belonging in the United States" (Fisherkeller 1997, 485–6, emphasis added). We need to come to grips with that reality, not keep it at arm's length. And at the same time, my own experience with the media deprivation exercise suggests to me that it is a mistake to conclude that all people, all the time, are in the vice-like grip of all media. The pervasive talk of "media saturation" overlooks the more complex reality, which is that people's attention is variable and selective; the deprivation exercise suggests that it is indeed very difficult for most of us to live without *some* media, but other media we can happily take or leave. Similarly, ethnographic research paints a more subtle and optimistic picture, showing people who engage enthusiastically with some messages, while letting much wash over them—and spending much of their time loving, caring and sparring with each other.

Thus, my intention in this book is not to suggest that we abandon the "audience" in despair, just because we cannot usually pin them down for clear study. And we do not need to abandon ethnography, just because as an enterprise it has become so complex. Even as we acknowledge the importance of global and national economic and political forces in constructing the mediascapes in which we live, we don't all need to become political economists. The "on the ground" perspective of the ethnographer is still crucial, and offers a dimension that no other approach can duplicate. We do need to move "beyond the audience" as a theoretically definable construct, but we should not be abandoning the goal of understanding real people, living real lives in which media play an ever-increasing, if certainly problematic, role. Above all, the ethnographic perspective on media reception offers a necessarily humanistic alternative to more controlled, scientific analyses of "media effects" or the "culture industry." Only ethnography can begin to answer questions about what people *really* do with media, rather than what we imagine they *might* do, or what close readings of texts *assume* they might do. As Jenson (1992) writes, ethnographers should never lose respect and empathy for the people with whom we work:

> They should not define people as collections of preferences to be analyzed and controlled, any more than they should define them as unwitting victims of ideology or advertising or media or mob mentalities or ego-fragmentation. Social inquiry can and should be a form of respectful engagement. It can and should illuminate the experiences of others *in their own terms,* because these 'others' are us, and human experiences intrinsically and inherently matter. (1992, 26, italics in original)

Over the years, I have worked with numerous research participants, who have allowed me to observe them, question them, and solicit their help in many different ways. Most have been enthusiastic and eager, becoming involved in the issues I raised, sometimes commenting on the finished product and offering me their continuing insights. I hope their voices come through this text as clearly as my own.

NOTES

Chapter 1

1. It is not my intention to review the history of audience study here. There are several excellent sources for such a review, including Alasuutari (1999), Moores (1992), Abercrombie and Longhurst (1998).

Chapter 2

1. For those with a short memory for scandal, Donna Rice was the woman with whom presidential hopeful Gary Hart was believed to be having the extra-marital affair that ended his candidacy; Marla Maples was the "other woman" credited with ending the marriage of mogul Donald Trump and his wife Ivana.
2. The importance of narrative as a way of structuring human experience has been recognized and explored across many disciplines in recent years. For some discussion of this, see, for example, Fisher 1987; Mechling 1991; Mitchell 1981.
3. Often these jokes combine scandalous figures; for example, the joke that Tonya Harding and Michael Jackson have opened a race track. She'll do the handicapping, while he rides the three-year-olds. Or the bartender who invented the Tonya-Bobbitt cocktail: a club soda with a slice, and so on.
4. A simple Internet search will yield multiple web sites devoted to jokes, parodies, and general lampooning of Clinton/Lewinsky. Some continue to be active in 2002, while others' most recent updates are several years old.

Chapter 3

1. The list members were aware of my research; although many members told me they would be willing to have their names used, I decided to maintain anonymity or use pseudonyms in the quotations. When one member is named in a message from another, I have substituted an initial to avoid unnecessary confusion.

Chapter 4

1. While the designation "Native American" is also widely used, I will use "American Indian," which is preferred in Minnesota.

2. Latest data show that in spite of revenue derived from gambling operations in some states, the overall living conditions for American Indians have declined, with unemployment over 50 percent on many reservations (Associated Press 2000).
3. I am indebted to Renee Botta and Carolyn Bronstein for sharing an unpublished paper using a similar methodology. In quoting transcripts, I do not attribute statements, except when quoting group exchanges, when initials are used to distinguish contributors.
4. Thousands of Indian children were forcibly removed from their families and educated in government-funded boarding schools, from the nineteenth century into the second half of the twentieth, with the goal of "cultural assimilation."

Chapter 6

1. I attempted to interview Xatasha Johnigan-Abdalla, the young woman who wrote the letter to *Ebony*, in order to shed some light on the circumstances behind her decision to write it. However, she refused to talk with me about the incident.
2. For a full discussion of the literature on news as story, see Bird & Dardenne (1988), Bird (1992a). For a personal account of how these processes operate among newspaper reporters, see Darnton (1975).
3. I wish to thank Dave Overton, News Director of KXAS-TV, Channel 5, Dallas/Fort Worth, who provided tapes of 12 stories on the CJ case aired between August 30 and October 21, 1991.
4. Young (1962) and di Leonardo (2000) point to the power of the erotically-charged "dusky maiden" stereotype in American and European culture from very early times to the twentieth century.

BIBLIOGRAPHY

Abercrombie, N. and B. Longhurst (1998). *Audiences.* London and Thousand Oaks, Calif.: Sage.

Abu-Lughod, L. (2000). Locating Ethnography. *Ethnography* 1(2): 261–67.

Aden, R. (1999). *Popular Stories and Promised Lands: Fan Cultures and Symbolic Pilgrimages.* Tuscaloosa: University of Alabama Press.

Agar, M. (1986). *Speaking of Ethnography.* Beverly Hills, Calif.: Sage.

———(1996). *The Professional Stranger: An Informal Introduction to Ethnography, 2nd Ed.* San Diego: Academic Press.

Alasuutari, P. (1999). Introduction: Three Phases of Reception Studies. In *Rethinking the Media Audience,* edited by P. Alasuutari. London: Sage.

Alderson, A. "Letter from New York: World Trade Centre Myth That Kept Hope Alive," <http://urbanlegends.about.com/gi/dynamic/offsite.htm?site=http://news.telegraph.co.uk/news/main.jhtml%3Fxml=/news/2001/09/23/wald23.xml>. *Daily Telegraph* (September 23 2001); Accessed online August 29, 2002.

Allor, M. (1988). Relocating the Site of the Audience. *Critical Studies in Mass Communication* 5: 217–33.

Altman, D. (1986). *AIDS in the Mind of America.* New York: Anchor Press/Doubleday.

Andersen, R. (1992). Oliver North and the News. In *Journalism and Popular Culture,* edited by P. Dahlgren and C. Sparks. Newbury Park, Calif.: Sage.

Ang, I. (1985). *Watching Dallas: Soap Opera and the Melodramatic Imagination.* London: Methuen.

———(1996). *Living Room Wars: Rethinking Media Audiences for a Postmodern World.* New York: Routledge.

Angrosino, M. V. (2002). *Doing Cultural Anthropology: Projects for Ethnographic Data Collection.* Prospect Heights, Ill.: Waveland.

Antoun, R. (1968). On the Significance of Names in an Arab Village. *Ethnology* 7: 158–70.

Appadurai, A. (1990). Disjuncture and Difference in the Global Cultural Economy. *Public Culture* 2: 1–24.

Askew, K. and R. R. Wilk, eds. (2002). *The Anthropology of Media: A Reader.* New York: Blackwell.

Ashley, J. and K. Jarratt-Ziemskie (1999). Superficiality and Bias: The (Mis)Treatment of Native Americans in U.S. Government Textbooks. *The American Indian Quarterly* 23(3–4): 49–62.

Associated Press (1991). AIDS Carrier Charged with Failing to Tell Sex Partner. *Dallas Times-Herald,* Oct. 3: A2.

———(2000). Casino Revenues Haven't Helped Most Indians. *St. Petersburg Times,* Sept. 3: A17.

Babbie, E. (1989). *The Practice of Social Research.* New York: Wadsworth.

Bacon-Smith, C. (1992). *Enterprising Women: Television Fandom and the Creation of Popular Myth.* Philadelphia: University of Pennsylvania Press.

Baird, R. (1996). Going Indian: Discovery, Adoption, and Renaming Toward a "True American," from *Deerslayer* to *Dances with Wolves.* In *Dressing in Feathers: The Construction of the Indian in American Popular Culture,* edited by S. E. Bird. Boulder, Colo.: Westview Press.

Balsamo, A. (1994). Feminism for the Incurably Informed. In *Flame Wars: The Discourse of Cyberculture,* edited by M. Dery. Raleigh, N.C.: Duke University Press.

Banks, M. (1998). Visual Anthropology: Image, Object and Interpretation. In *Image-based Research: A Sourcebook for Qualitative Researchers,* edited by J. Prosser. London: Falmer Press.

Barker, M. (1997). Taking the Extreme Case: Understanding a Fascist Fan of Judge Dredd. In *Trash Aesthetics: Popular Culture and its Audience,* edited by D.Cartmell, I. Q. Hunter, H. Keye, and I. Whelehan. London: Pluto Press.

———(1998). Critique: Audiences 'R' Us. In *Approaches to Audiences: A Reader,* edited by R. Dickinson, R. Harindranath, and O. Linné. London: Arnold.

Bascom, W. (1954). Four Functions of Folklore. *Journal of American Folklore* 67: 333–49.

Baym, N. K. (1997). Interpreting Soap Operas and Creating Community: Inside an Electronic Fan Culture. In *Culture of the Internet,* edited by S. Kiesler. Mahwah, N.J.: Lawrence Erlbaum.

———(2000). Tune in, Log on: Soaps, Fandom, and Online Community. Thousand Oaks, Calif.: Sage.

Beach, C. (1997). Recuperating the Aesthetic: Contemporary Approaches and the Case of Adorno. In *Beauty and the Critic: Aesthetics in the Age of Cultural Studies, edited by* J. Soderholm. Tuscaloosa: University of Alabama Press.

Becker, A. E., R. Burwell, S. E. Gilman, D. B. Herzog, and P. Hamburg (2002). Eating Behaviours and Attitudes Following Prolonged Exposure to Television among Ethnic Fijian Adolescent Girls. *British Journal of Psychiatry* 180: 509–514.

Behar, R. (1993). *Translated Woman: Crossing the Border with Esperanza's Story.* Boston: Beacon Press.

———(1996). The *Vulnerable Observer: Anthropology that Breaks Your Heart.* Boston: Beacon Press.

Bellah, R. N., R. Madsen, W. Sullivan, and S. Tipton (1996). *Habits of the Heart: Individualism and Commitment in American Life.* Berkeley: University of California Press.

Bennett, W. L. (1983). *News: The Politics of Illusion.* New York: Longman.

Berkhofer, R. (1979). *The White Man's Indian.* New York: Vintage Books.

Berkowitz, D. (1997). Non-Routine News and Newswork: Exploring a What-a-Story. In *Social Meanings of News: A Text Reader,* edited by D. Berkowitz. New York: Sage, 1997.

Bernard, H. R. ed. (1998). *Handbook of Methods in Cultural Anthropology.* Walnut Creek, Calif.: AltaMira Press.

Best, J. (1990). *Threatened Children: Rhetoric and Concern About Child-Victims.* Chicago: University of Chicago Press.

———(1991a). Bad Guys and Random Violence: Folklore and Media Constructions of Contemporary Deviants. *Contemporary Legend* 1: 107–21.

———(1991b). "Road-Warriors" on "Hair-Trigger Highways": Cultural Resources and the Media's Construction of the 1987 Freeway Shootings Problem. *Sociological Inquiry* 61: 327–45.

Bird, S. E. (1992a). *For Enquiring Minds: A Cultural Study of Supermarket Tabloids.* Knoxville: University of Tennessee Press.

———(1992b). Travels in Nowhere Land: Ethnography and the "Impossible" Audience. *Critical Studies in Mass Communication* 9: 250–60.

———(1994). Is that Me, Baby? Image, Authenticity, and the Career of Bruce Springsteen. *American Studies,* 35: 2, 39–58.

———(1996). Not My Fantasy: The Persistence of Indian Imagery in *Dr. Quinn, Medicine Woman.* In *Dressing in Feathers: The Construction of the Indian in American Popular Culture,* edited by S. E. Bird. Boulder, Colo.: Westview Press.

———(1999). "Gendered Representation of American Indians in Popular Media," *Journal of Communication* 49:3, 61–83.

———(2002). Taking It Personally: Supermarket Tabloids After September 11. In *Journalism After September 11,* edited by B. Zelizer and S. Allen. London: Routledge.

Bird, S. E. and J. Barber (2002). Constructing a Virtual Ethnography. In *Doing Cultural Anthropology: Projects for Ethnographic Data Collection,* edited by M. Angrosino. Prospect Heights, Ill.: Waveland.

Bird, S. E. and R. W. Dardenne (1988). Myth, Chronicle and Story: Exploring the Narrative Qualities of News. *Media, Myths and Narratives,* edited by J. W. Carey. Newbury Park: Sage.

Bird. S. E. and J. Jorgenson (2002). Extending the School Day: Gender, Class and the Incorporation of Technology in Everyday Life. In *Women and Everyday Uses of the Internet: Agency and Identity,* edited by M.Consalvo and S. Paasonen. New York: Peter Lang.

Bobo, J. (1995). *Black Women as Cultural Readers.* New York: Columbia University Press.

Bourdieu, P. (1984). *Distinction: A Social Critique of the Judgment of Taste,* translated by R. Nice. Cambridge, Mass.: Harvard University Press.

Brooks, P. (1994). Aesthetics and Ideology: What Happened to Poetics? *Critical Inquiry* 20(3): 509–523.

Brown, M. E. (1994). Soap Opera and Women's Talk: The Pleasure of Resistance. Thousand Oaks, Calif.: Sage.

Bruner, E. ed. (1984). *Text, Play, and Story: The Construction and Reconstruction of Self and Society.* Washington, D.C.: American Ethnological Society.

Brunsdon, C. (1990). Television: Aesthetics and Audiences. In *Logics of Television: Essays in Cultural Criticism,* edited by P. Mellencamp. Bloomington: Indiana University Press.

Burke, K. (1966). Language as Symbolic Action: Essays on Life, Literature, and Method. Berkeley: University of California Press.

Camitta, M. (1990). Gender and Method in Folklore Fieldwork. *Southern Folklore* 47(1): 21–31.

Campbell, J. (1949). *The Hero with a Thousand Faces.* New York: Pantheon.

Campbell, R. (1991). *60 Minutes: A Mythology for Middle America.* Urbana: University of Illinois Press.

Carey, J. W. (1986). The Dark Continent of American Journalism. In *Reading the News,* edited by R. K. Manoff and M. Schudson. New York: Pantheon.

———(1988). *Media, Myths and Narratives: Television and the Press.* Beverly Hills: Sage.

———(1989). *Communication as Culture: Essays on Media and Society.* Boston: Unwin Hyman.

Carroll, N. (1997). The Ontology of Mass Art. *The Journal of Aesthetics and Art Criticism* 55: 187–200.

Caughey, J. (1984). *Imaginary Social Worlds.* Lincoln: University of Nebraska Press.

———(1994). Gina as Steven: The Social and Cultural Dimensions of a Media Relationship. *Visual Anthropology Review* 10: 126–35.

Cawelti, J. (1976). *Adventure, Mystery, and Romance: Formula Stories as Art and Popular Culture.* Chicago: University of Chicago Press, 1976.

Chadwick, B., H. Bahr and S. Albrecht (1984). *Social Science Research Methods.* Englewood Cliffs, N.J.: Prentice-Hall.

Chaplin, E. (1998). Making Meanings in Art Worlds: A Sociological Account of the Career of John Constable and his Oeuvre, with Special Reference to "The Cornfield" (Homage to Howard Becker). In *Image-based Research: A Sourcebook for Qualitative Researchers,* edited by J. Prosser. London: Falmer Press.

Churchill, W. (1994). *Indians Are Us: Culture and Genocide in Native North America.* Monroe, Me.: Common Courage Press.

Clemmons, C. (1991). A Fearful Fallout from AIDS Letter to Magazine. *Dallas Times-Herald,* Sept. 3: C1, C5.

Clerc, S. (1996). Estrogen Brigades and "Big Tits" Threads: Media Fandom Online and Off. In *Wired Women: Gender and New Realities in Cyberspace,* edited by L. Cherny and E. R. Wise. Toronto: Seal Press.

Clifford, J. and G. Marcus (1986). *Writing Culture: The Poetics and Politics of Ethnography.* Berkeley: University of California Press.

Coates, N. (1998). Can't We Just Talk About Music: Rock and Gender on the Internet. In *Mapping the Beat: Popular Music and Contemporary Theory,* edited by T. Swiss, J. Sloop and A. Herman. Malden, Mass.: Blackwell.

Cohen, J. (1991). The "Relevance" of Cultural Identity in Audiences' Interpretations of Mass Media. *Critical Studies in Mass Communication* 5: 442–54.

Cohen, S. (1980). *Folk Devils and Moral Panics: The Creation of the Mods and Rockers.* New York: St. Martin's Press.

Cohen, T. (1999). High and Low Art, and High and Low Audiences. *The Journal of Aesthetics and Art Criticism* 57(2): 137–143.

Connell, I. (1992). Personalities in the Popular Media. In *Journalism and Popular Culture,* edited by P. Dahlgren and C. Sparks. Newbury Park: Sage.

Couldry, N. (2000). *The Place of Media Power: Pilgrims and Witnesses of the Media Age.* London: Routledge.

Crenson, M. and B. Ginsberg (2002). *Downsizing Democracy: How America Sidelined its Citizens and Privatized its Public.* Baltimore: Johns Hopkins University Press.

Curtis, P. (1997). Mudding: Social Phenomena in Text-Based Virtual Realities. In *Culture of the Internet,* edited by S. Kiesler. Mahwah, N.J.: Lawrence Erlbaum.

Dahl, D. (1998). In Race to be First, Truth Often Suffers. *St. Petersburg Times,* Jan. 31, 1A; 5A.

Dahlgren, P. (1992). Introduction. In *Journalism and Popular Culture,* edited by P. Dahlgren and C. Sparks. Newbury Park: Sage.

Dardenne, R. W. (1994). Student Musing on Life without Mass Media: Antidote for Silence, Boredom, and Thinking. *Journalism Educator* 49(3): 72–79.

Darnton, R. (1975). Writing News and Telling Stories. *Daedalus* 104: 175–94.

Davies, C. (1990). *Ethnic Humor from Around the World: A Comparative Analysis.* Bloomington: Indiana University Press.

Davies, S. (1999) Rock versus Classical Music. *Journal of Aesthetics and Art Criticism* 57: 193–204.

Davis, L. (1993). Protest Against the Use of Native American Mascots: A Challenge to Traditional American Identity. *Journal of Sport and Social Issues* 17(1): 9–22.

de Certeau, M. (1984). *The Practice of Everyday Life.* Berkeley: University of California Press.

Deggans, E. (1998). *Springer* Officials Decide Bestiality Episode is Unfit. *St. Petersburg Times,* May 23, B2.

Deloria, P. (1999). *Playing Indian.* New Haven: Yale University Press.

Dery, M. (1994). Introduction: Flame Wars. In *Flame Wars: The Discourse of Cyberculture,* edited by M. Dery. Raleigh, N.C.: Duke University Press.

di Leonardo, M. (2000). *Exotics at Home: Anthropologies, Others, American Modernity.* Urbana: University of Chicago Press.

DiGiovanna, J. (1996). Losing Your Voice on the Internet. In *High Noon on the Electronic*

Frontier: Conceptual Issues in Cyberspace, edited by P. Ludlow and M. Godwin. Cambridge, Mass.: MIT Press.

Doheny-Farina, S. (1996). *The Wired Neighborhood.* New Haven, Conn.: Yale University Press.

Drotner, K. (1994). Ethnographic Enigmas: The "Everyday" in Recent Media Studies. *Cultural Studies* 8(2): 341–57.

————(2000). Difference and Diversity: Trends in Young Danes' Media Cultures. *Media, Culture and Society* 22 (2): 149–66.

Dundes, A. and C. R. Pagter (1975). *Work Hard and You Shall Be Rewarded: Urban Folklore from the Paperwork Empire.* Bloomington: Indiana University Press.

Ebony Magazine (1991). The Ebony Advisor. September: 90.

Edwards, D. B. (1994). Afghanistan, Ethnography, and the New World Order. *Cultural Anthropology* 9(3): 345–60.

Eliade, M. (1958). *Patterns in Comparative Religion,* translated by R. Sheed. New York: Sheed and Ward.

Ellis, B. (1989). When Is a Legend: An Essay in Legend Morphology. In *The Questing Beast: Perspectives on Contemporary Legend,* edited by P. Smith and G. Bennett. Sheffield, U.K.: Sheffield Academic Press.

Ellis, C. and A. P. Bochner, eds. (1996). *Composing Ethnography: Alternative Forms of Qualitative Writing.* Walnut Creek, Calif.: AltaMira.

Emerson, R. M., R. I. Fretz, and L. L. Shaw (1995). *Writing Ethnographic Field Notes.* Chicago: University of Chicago Press.

Emke, I. (1992). Excising the Body (Politic): The AIDS Carrier Panic in Canada. Unpublished paper presented to International Communication Association, Miami.

Erni, J. (1989). Where Is the Audience? Discerning the (Impossible) Subject. *Journal of Communication Inquiry* 13(2): 30–42.

Evans, W. A. (1990). The Interpretive Turn in Media Research: Innovation, Iteration, or Illusion? *Critical Studies in Mass Communication* 7 (2): 147–68.

Fedler, F. (1989). *Media Hoaxes.* Ames: Iowa State University Press.

Fine, G. A. (1987). Welcome to the World of AIDS: Fantasies of Female Revenge. *Western Folklore* 46: 192–97.

Fisher, W. R. (1987). *Human Communication as Narration: Toward a Philosophy of Reason, Value, and Action.* Columbia: University of South Carolina Press.

————(1985). The Narrative Paradigm: In the Beginning. *Journal of Communication* 35: 74–89.

Fisherkeller, J. (1997). Everyday Learning about Identities among Young Adolescents in Television Culture. *Anthropology and Education Quarterly,* 28(4): 467–92

Fiske, J. (1988). Meaningful Moments. *Critical Studies in Mass Communication,* 5: 246–51.
(1989a). *Understanding Popular Culture.* Boston: Unwin Hyman.

————(1989b). *Reading the Popular.* Boston: Unwin Hyman.

————(1992). Popularity and the Politics of Information. In *Journalism and Popular Culture,* edited by P. Dahlgren and C. Sparks. Newbury Park: Sage.

Francis, D. (1992). T*he Imaginary Indian: The Image of the Indian in Canadian Culture.* Vancouver: Arsenal Pulp Press.

Frith, S. (1991). The Good, the Bad, and the Indifferent: Defending Popular Culture from the Populists. *Diacritics* 21(4): 102–15.

Fuller, M. and H. Jenkins (1995). Nintendo and New World Travel Writing: A Dialogue. In *Cybersociety: Commuter-Mediated Communication and Society,* edited by S. Jones. Thousand Oaks, Calif.: Sage.

Gadsen, N. L. (1994). HIV-Infected Prostitute Apprehended. *Sunday Patriot News.* March 13, Harrisburg, Penn.: B6.

Gajjala, R. (2002). An Interupted Postcolonial/Feminist Cyberethnography: Complicity and Resistance in the "Cyberfield." *Feminist Media Studies* 2(2): 177–94.

Gal, S. (1991). Between Speech and Silence: The Problematics of Research on Language and

Gender. In *Gender at the Crossroads of Knowledge: Feminist Anthropology in the Post-modern Era*, edited by M. di Leonardo. Berkeley: University of California Press.

Gans, H. (1979). *Deciding What's News*. New York: Pantheon.

Gillespie, M. (1995). *Television, Ethnicity, and Cultural Change*. London: Routledge.

Gilman, S. L. (1988). AIDS and Syphilis: The Iconography of Disease. *AIDS: Cultural Analysis, Cultural Activism*, edited by D. Crimp. Cambridge, Mass.: MIT Press.

Gilmore, D. (1987). *Aggression and Community: Paradoxes of Andalusian Culture*. New Haven, Conn.: Yale University Press.

Ginsburg, F. D. (1989). *Contested Lives: The Abortion Debate in an American Community*. Berkeley: University of California Press.

Ginsburg, F. D., L. Abu-Lughod, and B. Larkin, eds. (2002). *Media Worlds: Anthropology on New Terrain*. Berkeley: University of California Press.

Giroux, H. (2002). The Corporate War against Higher Education. *Workplace: A Journal for Academic Labor* 5(1), October. Accessed online Dec. 2, 2002, at <http://www.louisville.edu/journal/workplace>.

Gitlin, T. (2001). *Media Unlimited: How the Torrent of Images and Sounds Overwhelms Our Lives*. New York: Metropolitan Books.

Gluckman, M. (1963). Gossip and Scandal. *Current Anthropology* 4(3): 307–16.

Glynn, K. (1990). Tabloid Television's Transgressive Aesthetic: A Current Affair and the "Shows That Taste Forgot," *Wide Angle* 12(2): 22–44.

———(2000). *Tabloid Culture: Trash Taste, Popular Power, and the Transformation of American Television*. Durham, N.C.: Duke University Press.

Goffman, E. (1959). *The Presentation of Self in Everyday Life*. New York: Doubleday.

Goodwin, J. P. (1989). *More Man Than You'll Ever Be: Gay Folklore and Acculturation in Middle America*. Bloomington: Indiana University Press.

Goss, M. (1990). The Halifax Slasher and Other "Urban Maniac" Tales. *A Nest of Vipers: Perspectives on Contemporary Legend V*, edited by G. Bennett and P. Smith. Sheffield, U.K.: Sheffield Academic Press.

Gould, T. (1999). Pursuing the Popular. *Journal of Aesthetics and Art Criticism* 57(2): 120–135.

Goulet, J-G. (1994). A Narrative Ethnography of Experiences Among the Dene Tha. *Journal of Anthropological Research* 50(2): 113–39.

Graber, D. A. (1984). *Processing the News: How People Tame the Information Tide*. New York: Longman.

Gracyk, T. (1999). Valuating and Evaluating Popular Music. *Journal of Aesthetics and Art Criticism*, 57: 205–20.

Greenblatt, L. (2002). Days of Our Nights (Column). *Seattle Weekly*. August 1–7. Accessed online at <http://www.seattleweekly.com/features/0231/days-greenblatt.shtml>.

Grider, S. (1984). The Razor Blades in the Apple Syndrome. *Perspectives on Contemporary Legend*, edited by P. Smith. Sheffield, U.K.: Sheffield Academic Press.

Grindstaff, L.(1997). Producing Trash, Class, and the Money Shot: A Behind-the-Scenes Account of Daytime TV Talk Shows. In *Media Scandals: Morality and Desire in the Popular Culture Marketplace*, edited by J. Lull and S. Hinerman. Cambridge: Polity Press.

———(2002). *The Money Shot: Trash, Class, and the Making of TV Talk Shows*. Chicago: University of Chicago Press.

Gripsrud, J. (1992). The Aesthetics and Politics of Melodrama. In *Journalism and Popular Culture*, edited by P. Dahlgren and C. Sparks. Newbury Park: Sage.

———(1998). High Culture Revisited. In *Cultural Theory and Popular Culture: A Reader*, edited by J. Storey. Athens: University of Georgia Press.

Gritti, J. (1994). Rumors and *Faits Divers*: Paradoxical Affinities. *Foaftale News* 32: 4–5.

Grossberg, L. (1984a). Another Boring Day in Paradise: Rock and Roll and the Empowerment of Everyday Life. *Popular Music* 4: 225–57.

———(1984b). "I'd Rather Feel Bad than Not Feel Anything at All": Rock and Roll, Pleasure and Power. *Enclitic* 8: 94–111.

————(1988). Wandering Audiences, Nomadic Critics. *Cultural Studies* 2(3): 377–90.

————(1992). *We Gotta Get out of this Place: Popular Conservatism and Postmodern Culture.* London: Routledge.

————(1997). *Bringing it All Back Home: Essays on Cultural Studies.* Durham, N.C.: Duke University Press.

Grover, J. Z. (1988). AIDS: Keywords. In *AIDS: Cultural Analysis, Cultural Activism,* edited by D. Crimp. Cambridge, Mass.: MIT Press.

Gulyas, A. (2000). The Development of the Tabloid Press in Hungary. In *Tabloid Tales,* edited by C. Sparks and J. Tulloch. New York: Rowman and Littlefield.

Gulzow, M. and C. Mitchell (1980). "Vagina Dentata" and "Incurable Venereal Disease" Legends from the Vietnam War. *Western Folklore* 39: 306–17.

Gurak, L. (1997). *Persuasion and Privacy in Cyberspace: The Online Protests Over Lotus MarketPlace and the Clipper Chip.* New Haven, Conn.: Yale University Press.

Hahn, E. (1994). The Tongan Tradition of Going to the Movies. *Visual Anthropology Review* 10(1): 103–11.

Hall, S. (1981). Notes on Deconstructing "the Popular." In *People's History and Socialist Theory,* edited by R. Samuel. London: Routledge.

Hall, S. and P. Whannel (1964). *The Popular Arts.* London: Hutchinson Educational.

Hallin, D. (1992). The Passing of the "High Modernism" in American Journalism. *Journal of Communication* 42(3): 14–24.

————(2000). *La Nota Roja:* Popular Journalism and the Transition to Democracy in Mexico. In *Tabloid Tales,* edited by C. Sparks and J. Tulloch. New York: Rowman and Littlefield.

Hallin D. and T. Gitlin (1994). The Gulf War as Popular Culture and Television Drama. In *Taken by Storm: The Media, Public Opinion, and U.S. Foreign Policy in the Gulf War,* edited by W. L. Bennett and D. L. Paletz. Chicago: University of Chicago Press.

Hanson, J. R. and L. P. Rouse (1987). Dimensions of Native American Stereotyping. *American Indian Culture and Research Journal* 11(4): 33–58.

Hardy, S. and R. Kukla (1999). A Paramount Narrative: Exploring Space on the Starship Enterprise. *Journal of Aesthetics and Art Criticism* 57: 177–91.

Hartley, J. (1992). *The Politics of Pictures: The Creation of the Public in the Age of Popular Media.* London: Routledge.

Harvard Business School & Kennedy School of Government (1998). Business and the Internet, course website. Accessed online Nov. 10, 2002, at <http://www.ksg.harvard.edu/iip/stp307>.

Haskins, J. (1984). Morbid Curiosity and the Mass Media: A Synergistic Relationship. In *Proceedings of the Conference on Morbid Curiosity and the Mass Media,* edited by J. Crook, J. Haskins and P. Ashdown. Knoxville: University of Tennessee.

Hayward, J. (1997). *Consuming Pleasures: Active Audiences and Serial Fictions from Dickens to Soap Opera.* Lexington: University Press of Kentucky.

Heider, D. (1998). Elephants and Green Beans: Propping Up Local TV News. *Chronicle of Higher Education,* XLIV: 37, May 22, B8.

Higgins, K and J. Rudinow, J. (1999). Introduction to Special Issue on Aesthetics and Popular Culture. *The Journal of Aesthetics and Art Criticism* 57(2): 109–18.

Hills, M. (2002). *Fan Cultures.* London: Routledge.

Hobson, D. (1982). *Crossroads: The Drama of a Soap Opera.* London : Methuen.

Hogshire, J. (1989). Boy, 17, is Married to 67-year-old Granny... Who Hasn't Aged in Years! *National Examiner,* Dec. 19: 23.

————(1992). Personal telephone interview, March 13.

Hopper, R. (1992). *Telephone Conversation.* Bloomington: Indiana University Press.

Humdog (1996). Pandora's Vox. In *High Noon on the Electronic Frontier: Conceptual Issues in Cyberspace,* edited by P. Ludlow and M. Godwin. Cambridge, Mass.: MIT Press.

Hunter, I. (1992). Aesthetics and Cultural Studies. In *Cultural Studies,* edited by L. Grossberg, C. Nelson and P. Treichler. New York: Routledge.

Jenkins, H. (1992). *Textual Poachers: Television Fans and Participatory Culture.* New York: Routledge.

Jensen, K. B. (1990). The Politics of Polysemy: Television News, Everyday Consciousness, and Political Action. *Media, Culture, and Society,* 12(1): 57–77.

Jenson, J. (1992). Fandom as Pathology: The Consequences of Characterization. In *The Adoring Audience: Fan Culture and Popular Media,* edited by L. Lewis. New York: Routledge.

Johnson, R. (1983). *What is Cultural Studies Anyway?* Birmingham, U.K.: Centre for Contemporary Cultural Studies General Series SP74.

Jones, S. G. (1998). *Doing Internet Research: Critical Issues and Methods for Examining the Net.* Thousand Oaks, Calif.: Sage.

Keesing, R. (1985). Kwaio Women Speak: The Micropolitics of Autobiography in a Solomon Island Society. *American Anthropologist* 87(1): 27–39.

Kellner, D. (1999). *The X-Files* and the Aesthetics and Politics of Postmodern Pop. *The Journal of Aesthetics and Art Criticism* 57(2): 161–75.

Kling, A. (2002). Is Blogging a Fad? *Corante: Tech News.* Accessed online Dec. 3, 2002, at <http://www.corante.com/bottomline/articles/20020621-875.shtml>.

Knight, D. (1999). Why We Enjoy Condemning Sentimentality: A Meta-Aesthetic Perspective. *The Journal of Aesthetics and Art Criticism* 57(4): 411–20.

Koplowitz, H. (1988). Cyber Monica Lewinsky Webpage. Accessed online Oct. 16, 2002, at <http://onyx.he.net/~hotmoves/LIC/monica.html>.

Kottak, C. P. (1990). *Prime-time Society: An Anthropological Analysis of Television and Culture.* Belmont, Calif.: Wadsworth.

Krajicek, D. J. (1998). *Scooped: Media Miss Real Story on Crime While Chasing Sex, Sleaze, and Celebrities.* New York: Columbia University Press.

Kramarae, C. (1995). Backstage Critique of Virtual Reality. In *Cybersociety: Commuter-Mediated Communication and Society,* edited by S. G. Jones. Thousand Oaks, Calif.: Sage.

Krepcho, M., M. Smerick, A. Freeman, and X. X. Alfaro (1993). Harnessing the Energy of the Mass Media: HIV Awareness in Dallas. *American Journal of Public Health* 83: 283–85.

Lamas, D. (2002). Bridal Bargains: Planning a Wedding Can Tax Anyone's Budget. *Miami Herald,* Jun. 17, 2002. Accessed online Nov. 10, 2002, at <http://www.bayarea.com/mld/bayarea/living/occasions/3489494.htm>.

Langellier, K. and D. Hall. (1989). Interviewing Women: A Phenomenological Approach to Feminist Communication Research. In *Doing Research on Women's Communication: Perspectives on Theory and Method,* edited by K. Carter and C. Spitzack. Norwood, N.J.: Ablex.

Langer, J. (1992). Truly Awful News on Television. In *Journalism and Popular Culture,* edited by P. Dahlgren and C. Sparks. Newbury Park: Sage.

———(1998) *Tabloid Television: Popular Journalism and the "Other News."* London: Routledge.

Lawrence, J. S. and R. Jewett (2002). *The Myth of the American Superhero.* New York: Eerdmans.

Lewis, C. (1997). Hegemony in the Ideal: Wedding Photography, Consumerism, and Patriarchy. *Womens Studies in Communication* 20(2): 167–87.

Lewis, L. (1992). Introduction. In *The Adoring Audience: Fan Culture and Popular Media,* edited by L. Lewis. New York: Routledge.

Lind, R. (1996). Diverse Interpretations: The "Relevance" of Race in the Construction of Meaning in, and the Evaluation of, a Television News Story. *Howard Journal of Communications* 7: 53–74.

Livingstone, S. (1998). Audience Research at the Crossroads: The "Implied Audience" in Media and Cultural Theory. *European Journal of Cultural Studies* 1(2): 193–217.

———(1999). Imaginary Spaces: Television, Technology and Everyday Consciousness. In *Television and Common Knowledge,* edited by J. Gripsrud. London: Routledge.

Lule, J. (2001). *Daily News, Eternal Stories: The Mythological Role of Journalism.* New York: Guilford.

Lull, J. (1990). Inside *Family Viewing: Ethnographic Research on Television's Audience.* New York: Routledge.

Lyons, J. and J. Needham (1991). CJ's Vendetta Heightens Fear and Awareness of AIDS. *Dallas Times-Herald,* Sept. 3: AI, A8.

MacDonald, D. (1957/1998). A Theory of Mass Culture. In *Cultural Theory and Popular Culture: A Reader,* edited by J. Storey. Athens: University of Georgia Press.

Malinowski, B. (1954). *Magic, Science and Religion, and Other Essays.* Garden City, N.Y.: Doubleday.

Maltz, D. and R. Borker (1999). A Cultural Approach to Male-Female Miscommunication. In *Applying Cultural Anthropology,* edited by A. Podolefsky and P. J. Brown. Mountain View, Calif.: Mayfield.

Mankekar, P. (1999). *Screening Culture, Viewing Politics: An Ethnography of Television, Womanhood, and Nation in Postcolonial India.* Durham, N.C.: Duke University Press.

Marcus, G. (1998). Ethnography through Thick and Thin. Princeton: Princeton University Press.

———(1986). Contemporary Problems of Ethnography in the Modern World System. In *Writing Culture: The Poetics and Politics of Ethnography,* edited by J. Clifford and G. Marcus. Berkeley: University of California Press.

Martin, E. (1994). *Flexible Bodies: Tracking Immunity in American Culture from the Days of Polio to the Age of AIDS.* Boston: Beacon.

Martin-Barbero, J. (1993). *Communication, Culture, and Hegemony: From the Media to Mediations,* translated by E. Fox and R. White. Newbury Park: Sage.

Martinez, W. (1992). Who Constructs Anthropological Knowledge? Toward a Theory of Ethnographic Film Spectatorship. In *Film as Ethnography,* edited by P. I. Crawford and D. Turton. Manchester: Manchester University Press.

———(1996). Deconstructing the "Viewer": From Ethnography of the Visual to Critique of the Occult. In *The Construction of the Viewer: Media Ethnography and the Anthropology of Audiences,* edited by P. I. Crawford and S. B. Hafsteinnson. Aarhus, Denmark: Intervention Press.

Mastrolia, B. A. (1997). The Media Deprivation Experience: Revealing Mass Media as both Message and Massage. *Communication Education* 46: 203–10.

McChesney, R. W. (2000). Rich Media, Poor Democracy: Communication Politics in Dubious Times. New York: The New Press.

McFarland, J. (1985). Those Scribbling Women: A Cultural Study of Mid-Nineteenth Century Romances for Women. *Journal of Communication Inquiry* 9: 33–53.

McGuigan, J. (1998). Trajectories of Cultural Populism. In *Cultural Theory and Popular Culture: A Reader,* edited by J. Storey. Athens: University of Georgia Press.

McKee, A. (2001). Which is the Best *Doctor Who* Story? A Case Study in Value Judgements outside the Academy. *Intensities: The Journal of Cult Media* 1. Accessed online Oct. 5, 2002 at <http://www.cult-media.com/issue1/Amckee.htm>.

McKinley, E. G. (1997). *Beverly Hills, 90210: Television, Gender, and Identity.* Philadelphia: University of Pennsylvania Press.

McRobbie, A. (1982). The Politics of Feminist Research: Between Talk, Text and Action. *Feminist Review* 12: 46–57.

McTavish, D. G. and H. J. Loether (2002). *Social Research: An Evolving Process.* Boston: Allyn and Bacon.

Mechling, J. (1991). Homo Narrans Across the Disciplines. *Western Folklore* 50: 41–52.

Meehan, E. R. (1990). Why We Don't Count: The Commodity Audience Logics of Television. In *Essays in Cultural Criticism,* edited by P. Mellencamp. Bloomington: Indiana University Press.

Mellencamp, P. (1992). *High Anxiety: Catastrophe, Scandal, Age, and Comedy.* Bloomington: Indiana University Press.

Meyrowitz, J. (1985). *No Sense of Place: The Impact of Electronic Media on Social Behavior.* New York: Oxford University Press.

Mikkelson, B. (2000). The AIDS Mary Legend. Aug. 29. Accessed online Oct. 18, 2002 at <http://www.snopes.com/horrors/madmen/aidsmary2000>.

Miller, D. (1992). *The Young and the Restless* in Trinidad: A Case of the Local and the Global in Mass Consumption. In *Consuming Technology,* edited by R. Silverstone and E. Hirsch. London: Routledge.

Miller. D. and D. Slater (2000). *The Internet: An Ethnographic Approach.* Oxford: Berg.

Mills, M. (1990). Critical Theory and the Folklorists: Performance, Interpretive Authority, and Gender. *Southern Folklore* 47(1): 5–15.

Mintz, S. W. (1985). *Sweetness and Power.* New York: Viking, 1985.

Mitchell, W. E. (1981). *On Narrative.* Chicago: University of Chicago Press.

Moeller, S. D. (1998). *Compassion Fatigue: How the Media Sell Disease, Famine, War, and Death.* London: Routledge.

Moores, S. (1993). *Interpreting Audiences: The Ethnography of Media Consumption.* London: Sage.

Morley, D. (1980). *The Nationwide Audience.* London: British Film Institute.

———(1986). *Family Television: Cultural Power and Domestic Leisure.* London: Comedia.

———(1999). "To Boldly Go . . . " The "Third Generation" of Reception Studies. In *Rethinking the Media Audience,* edited by P. Alasuutari. London: Sage.

Morley, D. and R. Silverstone (1991). Communication and Context: Ethnographic Perspectives on the Media Audience. In *A Handbook of Qualitative Methodologies for Mass Communication Research,* edited by K. Jensen and N. Jankowski. London: Routledge.

Morse, R. (1999). The Age of Passing Acquaintance. *St. Petersburg Times,* May 22: 15A.

Morris, M. (1988). Banality in Cultural Studies. *Discourse* 10(2): 3–29.

Morris, N. (2002). The Myth of Unadulterated Culture Meets the Threat of Imported Media. *Media, Culture and Society* 24: 278–89.

Murad, M. (2002). Shouting at the Crocodile. In *Into the Buzzsaw: Leading Journalists Expose the Myth of the Free Press,* edited by K. Borjesson. New York: Prometheus.

Murdock, G. (1997). Thin Descriptions: Questions of Method in Cultural Analysis. In *Cultural Methodologies,* edited by J. McGuigan. London: Sage.

Newcomb, H., ed. (1974). *TV: The Most Popular Art.* Garden City, N.Y.: Anchor.

Newport, F. (1996). Younger Adults Up on Popular Culture and Sports Personalities, but Weak on Political Figures. Gallup Poll Archives, 6:11. The Gallup Organization, Princeton. Accessed online Nov. 7, 2001 at <http://www.gallup.com/poll_archives>.

Northrup, J. (1995). Indian Issues Column, *Duluth News Tribune,* July 26, 1995: 7A.

Oring, E. (1987). Jokes and the Discourse of Disaster. *Journal of American Folklore* 100: 276–86.

———(1990). Legend, Truth, and News. *Southern Folklore* 47: 163–77.

Owen, R. (1999). Gone but not Forgotten: When a Favorite TV Show is Canceled, Viewers Fight to Save It. *Pittsburgh Post-Gazette.* April 4. Accessed online April 10 at <http://www.post-gazette.com/magazine/ 19990404attach4.asp>.

O'Connor, J. J. (1993). It's Jane Seymour, M.D., in the Wild and Wooly West. *New York Times,* Feb. 4: B5.

Parameswaran, R. (1999). Western Romance Fiction as English-Language Media in Postcolonial India. *Journal of Communication* 49(3): 84–105.

Parker, K. and C. Deane (1997). Ten Years of the Pew News Interest Index: Report of Presentation at the 1997 meeting of the American Association for Public Opinion Research. Accessed online Nov. 7, 2001 at <http://www.people-press.org/index.htm>.

Peterson, J. (2002). Breaking Down Peter Chernin's Comdex Keynote. Nov. 23. Accessed online Dec. 2, 2002 at <http://www.way.nu/archives/000493.html#000493>.

Pew Center for Research on People and the Press (1998). White House Scandal Has Families Talking. Sept. 30. Accessed online Nov. 1, 2002 at <http://people-press.org/reports/display.php3?ReportID=781998>.

Postman, N. (1985). *Amusing Ourselves to Death: Public Discourse in the Age of Show Business.* New York: Viking, 1985.

Precker, M. (1991). CJ's Story Draws National Attention. *Dallas Morning News,* Oct. 12: IC, 3C.

———(1993). Personal telephone interview, April 27.

Press, A. (1991). *Women Watching Television: Gender, Class, and Generation in the American Television Experience.* Philadelphia: University of Pennsylvania Press.

Price, V. and E. Czilli (1996). Modeling Patterns of News Recognition and Recall. *Journal of Communication* 46(2): 55–78.

Putnam, R. D. (1996). The Strange Disappearance of Civic America. *The American Prospect* (Winter): 34–48.

Quam, M. D. (1990). The Sick Role, Stigma, and Pollution. In *Culture and AIDS,* edited by D. A. Feldman. New York: Praeger.

Quirk, P. (2000). Scandal Time: The Clinton Impeachment and the Distraction of American Politics. In *The Clinton Scandal and the Future of American Government,* edited by M. J. Rozell and C. Wilcox. Washington, D.C.: Georgetown University Press.

Radway, J. (1984). *Reading the Romance: Women, Patriarchy, and Popular Literature.* Philadelphia: University of Pennsylvania Press.

———(1988). Reception Study: Ethnography and the Problems of Dispersed Audiences and Nomadic Subjects. *Cultural Studies* 2(3): 359–76.

Rakow, L. (1992). *Gender on the Line: Women, the Telephone, and Community Life.* Urbana: University of Illinois Press.

Redvern-Vance, N. (1999). Narratives of Women Veterans: The Experience of Sexual Abuse. Unpublished Ph.D. dissertation, Department of Anthropology, University of South Florida.

Rheingold, H. (1996). A Slice of My Life in My Virtual Community. In *High Noon on the Electronic Frontier: Conceptual Issues in Cyberspace,* edited by P. Ludlow and M. Godwin. Cambridge, Mass.: MIT Press.

Robinson, T. N., Chang, J. Y., Haydel, K. F., and J. D. Killen (2001). Overweight Concerns and Body Dissatisfaction Among Third-Grade Children. *Journal of Pediatrics,* 138(2): 181–87.

Russell, R. (1999). Make Believe Indian. *New Times Los Angeles Online,* April 8–14.

Sarch, A. (1993). Making the Connection: Single Women's Use of the Telephone in Dating Relationships with Men. *Journal of Communication* 43(2): 128–44.

Schroeder, F. E. H. (1982). National Enquirer is National Fetish! The Untold Story! In *Objects of Special Devotion: Fetishes and Fetishism in Popular Culture,* edited by R. B. Browne. Bowling Green: Popular Press.

Seiter, E. (1999). *Television and New Media Audiences.* Oxford: Oxford University Press.

Shilts, R. (1987). *And the Band Played On: Politics, People, and the AIDS Epidemic.* New York: St. Martins.

Shiveley, J. (1992). Cowboys and Indians: Perceptions of Western Films Among American Indians and Anglos. *American Sociological Review* 57: 725–34.

Shusterman, R. (1999). Moving Truth: Affect and Authenticity in Country Musicals. *Journal of Aesthetics and Art Criticism* 57(2): 221–33.

Siikala, A-M. (1984). The Praxis of Folk Narratives. In *ARV: Scandinavian Yearbook of Folklore,* edited by B. R. Jonsson. Stockholm: Almquist and Wiksell.

Smith, P. (1990). "AIDS: Don't Die of Ignorance": Exploring the Cultural Complex. In *A Nest of Vipers: Perspectives on Contemporary Legend V,* edited by G. Bennett and P. Smith. Sheffield, U.K.: Sheffield Academic Press.

————(1992). "Read All About It! Elvis Eaten by Drug-Crazed Giant Alligators": Contemporary Legend and the Popular Press. *Contemporary Legend* 2: 41–70.

Sonner, M. and C. Wilcox (1999). Forgiving and Forgetting: Public Support for Bill Clinton During the Lewinsky Scandal. *Political Science and Politics: PS.* September: 554–57.

Sontag, S. (1989). *AIDS and Its Metaphors.* New York: Farrar, Strauss and Giroux.

Sparks, C. (1992). Popular Journalism: Theories and Practice. In *Journalism and Popular Culture,* edited by P. Dahlgren and C. Sparks. London: Sage.

————(2000). Introduction: The Panic over Tabloid News. In *Tabloid Tales,* edited by C. Sparks and J. Tulloch. New York: Rowman and Littlefield.

Sparks, C. and Tulloch, J., eds. (2000). *Tabloid Tales.* New York: Rowman and Littlefield.

Speer, J. H. (1992). Folklore Studies and Communication Studies: Shared Agendas. *Southern Folklore,* 49 (1): 5–18.

Spitulnik, D. (1993). Anthropology and Mass Media. *Annual Review of Anthropology* 22: 293–315.

Stephens, M. (1988). *History of News: From the Drum to the Satellite.* New York: Viking.

Stice, E., C. Hayward, R. Cameron, J. Killen, and B. Taylor (2000). Body-image and Eating Disturbances Predict Onset of Depression among Female Adolescents: A Longitudinal Study. *Journal of Abnormal Psychology* 109 (3): 438–44.

Stoll, C. (1995). *Silicon Snake Oil.* New York: Doubleday.

Sullivan, A. (2002). The Blogging Revolution: Weblogs Are to Words What Napster Was to Music. *Wired* Magazine, Issue 10.05, May. Accessed online Nov. 22, 2002, at <http://www.wired.com/wired/archive/10.05/mustread.html?pg=2>.

Tamarkin, B. (1993). *Rumor Has It: A Curio of Lies, Hoaxes, and Hearsay.* New York: Prentice Hall.

Tannen, D. (1993). *Gender and Conversational Interaction.* New York: Oxford University Press.

Taylor, A. (1996). Cultural Heritage in Northern Exposure. In *Dressing in Feathers: The Construction of the Indian in American Popular Culture,* edited by S. E. Bird. Boulder, Colo.: Westview Press.

Taylor, C. C. (1990). AIDS and the Pathogenesis of Metaphor. In *Culture and AIDS,* edited by D. A. Feldman. New York: Praeger.

Third World and Environment Broadcasting Project (2000). Losing Perspective: Global Affairs on British Terrestrial Television, 1989-1999. Details accessed online August 1, 2002 at <http://www.culture.gov.uk/creative/dti-dcms_3we.PDF>.

Thompson, J. (1990). *Ideology and Modern Culture: Critical Social Theory in the Era of Mass Communication.* Cambridge, U.K.: Polity Press.

Tomlinson, J. (1997). And Besides, the Wench Is Dead: Media Scandals and the Globalisation of Communication. In *Media Scandals: Morality and Desire in the Popular Culture Marketplace,* edited by J. Lull and S. Hinerman. London: Polity Press.

Treichler, P. A. (1988). AIDS, Gender, and Biomedical Discourse. In *AIDS, the Burden of History,* edited by E. Fee and D. M. Fox. Berkeley: University of California Press.

Trejo, J. (1979). Coyote Tales: A Paiute Commentary. In *Readings in American Folklore,* edited by J. H. Brunvand. New York: W. W. Norton.

Tompkins, J. P. (1992). *West of Everything: The Inner Life of Westerns.* New York : Oxford University Press.

Turkle, S. (1997). Constructions and Reconstructions of Self in Virtual Reality: Playing in the Muds. In *Culture of the Internet,* edited by S. Kiesler. Mahwah, N. J.: Erlbaum.

Turner, P. A. (1993). *I Heard It Through the Grapevine: Rumor in African-American Culture.* Berkeley: University of California Press.

Turner, V. (1985). *On the Edge of the Bush: Anthropology as Experience,* edited by E. L. B. Turner. Tucson: University of Arizona Press.

Twitchell, J. B. (1992). *Carnival Culture: The Trashing of Taste in America.* New York: Columbia University Press.

Van Zoonen, L. (1994). *Feminist Media Studies*. Thousand Oaks, Calif.: Sage.

Victor, J. S. (1993). *Satanic Panic: The Creation of a Contemporary Legend*. Chicago: Open Court Press.

Warnick, B. (1999). Masculinizing the Feminine: Inviting Women on Line Circa 1997. *Critical Studies in Mass Communication* 16: 1–19.

Wester, F. and N. Jankowski (1991). The Qualitative Tradition in Social Science Inquiry: Contributions to Mass Communication Research. In *A Handbook of Qualitative Methodologies for Mass Communication Research*, edited by K. Jensen and N. Jankowski. London: Routledge.

Whitehead, A. N. (1943/1954). *Dialogues*, as recorded by Lucien Price, June 10. Boston: Little, Brown.

Whitt, L. (1995). Cultural Imperialism and the Marketing of Native America. *American Indian Culture and Research Journal* 19: 1–32.

Wilk, R. R. (1994). Colonial Time and TV Time: Television and Temporality in Belize. *Visual Anthropology Review* 10(1): 94–111.

Williams, J. (2001). The Personal is Political: Thinking Through the Clinton/Lewinsky/Starr Affair. *Political Science and Politics: PS*. March: 93–98.

Williams, R. (1961). *The Long Revolution*. London: Chatto and Windus.

Williamson, J. (1986). The Problems of Being Popular. *New Socialist*, September: 14–19.

Willis, P. (1998). Notes on Common Culture: Towards a Grounded Aesthetics. *European Journal of Cultural Studies* 1(2): 163–76.

Woodward, K. (1997). *Identity and Difference*. London: Sage.

Worth, D. (1990). Minority Women and AIDS: Culture, Race, and Gender. In *Culture and AIDS*, edited by D. A. Feldman. New York: Praeger.

Young, P. (1962). The Mother of Us All: Pocahontas Reconsidered. *Kenyon Review* 24: 391–415.

Zelizer, B. and S. Allan, eds. (2002). *Journalism After September 11*. London: Routledge.

Zoglin, R. (1993). Frontier Feminist. *Time*, March 1: 63–64.

Zuckerman, M. (1984). Is Curiosity about Morbid Events an Expression of Sensation Seeking. In *Proceedings of the Conference on Morbid Curiosity and the Mass Media*, edited by J. Crook, J. Haskins and P. Ashdown. Knoxville: University of Tennessee Press.

INDEX